ЛИТЕЙНЫЕ СВОЙСТВА МЕТАЛЛОВ И СПЛАВОВ

LITEINYE SVOISTVA METALLOV I SPLAVOV

CASTING PROPERTIES OF METALS AND ALLOYS

CASTING PROPERTIES
OF METALS AND ALLOYS

by

A. M. Korol'kov

Authorized translation from the Russian

Springer Science+Business Media, LLC

Library of Congress Catalog Card Number 61-18757

The Russian text was published by the USSR Academy of Sciences Press
for the A. A. Baikov Institute of Metallurgy
in Moscow in 1960

А. М. Корольков

Литейные свойства металлов и сплавов

ISBN 978-1-4899-4676-8   ISBN 978-1-4899-4674-4 (eBook)
DOI 10.1007/978-1-4899-4674-4

CONTENTS

INTRODUCTION

The properties of castings and the quality of the metal in ingots are largely dependent on the charge, melting conditions in furnaces of different constructions, treatment of the metal in the furnaces by liquid slags and gases, modification, casting and solidification conditions, heat treatment, and so forth. The quality of castings is definitely influenced by the processes of preparing the liquid metal, and its casting and solidification in the molds. These processes are so interrelated that even a well-prepared liquid metal may be spoiled in the casting process or during solidification if the conditions are unsuitable.

Under industrial conditions, the course of these processes or their termination is assessed mainly from the results of the analysis of the composition of the alloy, from fracture tests and best of all from tests of the degree of saturation of the liquid metal by gases. Examination of the properties of the liquid metal is carried out only in isolated cases and is far from being adequate. This is due to the fact that, heretofore, not many relationships have been established between the quality of the final castings and the characteristic properties of the liquid metal, such as: viscosity, surface tension, degree of saturation by gases and also its properties in the liquid-solid state (in the casting and crystallization period): fluidity, oxidizability, liberation of gas, shrinkage, etc. It has so far not been found possible to determine these properties for complex industrial alloys on the basis, for example, of the individual properties of the alloy components. At the same time, the results of individual investigations cannot always be used in explanation of the observed phenomena and variations in the actual industrial processes. A characteristic example of this is provided by contradictions in regard to the applicability of the laws of crystallization associated with supercooling (Tammann's well-known curves) to the crystallization of castings of different weights. Views on the hereditary properties of alloys are still divided, although this question has been with us for half a century.

The study of foundry processes on a scientific basis, which was commenced in the middle of the last century by P. P. Anosov, A. S. Lavrov, and N. V. Kalakutskii, made it possible to pass from a mere accumulation of facts to an examination of the relationship between crystallization, the formation and development of shrinkage defects, and segregation.

P. P. Anosov (1797 - 1851) was the first to point out the regular manner in which the properties of steel depend on its composition and structure and on the conditions of its solidification; he laid down the foundations of macroscopic and microscopic methods in the analysis of steel, long before they came to be employed in other countries [3, 4]. The farsightedness of this scientist enabled him to point out the most important methods of pouring sound ingots of steel and to take note of such properties of metals as fluidity, shrinkage, etc.

In the work of P. P. Anosov we find the first steps of investigations in the field of crystallization, structure, and properties of steel, which were developed in the subsequent work of Russian metallurgists.

The direct successors of P. P. Anosov in the study of the processes of casting and crystallization of metals were N. V. Kalakutskii (1831 - 1899) and A. S. Lavrov (1838 - 1904).

N. V. Kalakutskii, employed as artillery inspector at the Zlatoust and Obukh Arsenals, made a study of the properties of steel in relation to casting, crystallization, and subsequent forging conditions; he associated the quality of the steel in these production stages with its structure. The great merit of N. V. Kalakutskii is his discovery (made with A. S. Lavrov) of segregation in steel. In studying the crystallization process of steel, N. V. Kalakutskii was the first to discover the connection between the composition of steel and its solidification time in crucibles. Thus, he found that mild steel solidified in the course of 3 min, medium steel in 7 - 8 min, and the hardest steel in 10 min [6, p. 32 ff]. The best-known work of N. V. Kalakutskii is that on shrinkage phenomena and the residual stresses produced in ingots during solidification and subsequent cooling.

N. V. Kalakutskii pointed out the real source and mechanism of the production of internal stresses in ingots due to shrinkage in relation to the subsequent course of their solidification and cooling. He developed a method of determining the internal stresses in a metal, which, like many other results of his investigations, retains its significance even today. Descriptions of the work of N. V. Kalakutskii and its later development are to be found in modern textbooks on foundry practice [7, 8].

In the work of A. S. Lavrov, we find the solution of a number of vital problems concerning the theory and practice of the casting and crystallization of metals. Lavrov was the first to describe the character and regular distribution of shrinkage cavities in steel ingots and to indicate steps for combating them. In regard to contemporary problems, the great service of A. S. Lavrov lies in the method he proposed for casting ingots in molds immersed in water. Unfortunately, this side of the scientific and engineering activities of A. S. Lavrov has so far remained undeservedly forgotten and is not mentioned in any work on semicontinuous and continuous casting methods, using machines of various constructions. In the meantime, A. S. Lavrov succeeded in producing cast steel of almost the same density as forged steel by eliminating porosity and locating the shrinkage cavity or pipe entirely in the upper part of the ingot.

The importance of the work of N. V. Kalakutskii and A. S. Lavrov in the development of the science of metals was highly appreciated by D. K. Chernov. "Our literature must be proud of the work of Lavrov and Kalakutskii: they were the first to point out the distribution of cavities in steel ingots and their dependence on melting and casting conditions, the distribution of the density of the steel itself in different parts of the ingots and the heterogeneity of its chemical composition . . ."[14, p. 86].

The flourishing science of metals and the struggle for quality in cast steel in the last century are associated with the period of activity of the great Russian scientist D. K. Chernov (1839 - 1921), whose work provided a sure basis for scientific metallurgy. His discovery of the phenomenon of thermal transformations in steel, and the discovery of the well-known Chernov critical points were brilliant achievements of Russian scientific thinking. His classic investigations on the crystallization and structure of steel ingots gave rise to an advanced theory of the crystallization of metals, which even today has not lost its significance. The pages of the scientific and technical literature on metals still contain references to the question of crystallization centers, first raised by D. K. Chernov long before the appearance of Tammann's work.

In an article "Investigations relating to the structure of cast steel ingots" (1878), D. K. Chernov advocated the wider use of casting processes instead of the comparatively costly process of forging. "Since the iron industry now possesses the means of producing steel in liquid form, in every possible quality, in large amounts and by cheap methods, a direct consequence of such progress ought to be the application of foundry practice to the production of all kinds of steel articles" [10]. The fundamental scientific content of this paper aimed at establishing the laws governing the crystallization of steel in the ingot, segregation phenomena, shrinkage, liberation of gas, and the formation of "discontinuous" crystals (dendrites). Recording the centenary of the birth of D. K. Chernov, Academician A. A. Baikov said that D. K. Chernov "was the first to point out quite convincingly and definitely that the crystals are the result of the simultaneous crystallization of iron and carbon...," i.e., they represent solutions of iron and carbon; only he did not call the crystals "solid solutions," since that expression did not appear in scientific literature until 15 years later (Van t'Hoff)"... "The importance of D. K. Chernov in regard to metallurgy may be compared with that of D. I. Mendeleev in regard to chemistry" [16, pp. 255 and 264].

The work of P. P. Anosov, A. S. Lavrov, D. K. Chernov and others brought Russian metallurgy into a leading position in the world. The famous German scientist E. Hein, at the beginning of the 20th century, wrote that the study of metals was not so well developed in Germany as in Russia [15].

The long road to the perfection of casting processes reflects the considerable difficulties arising from the poor casting properties of alloys. This applies in particular to alloys which solidify over a wide temperature range. The search for new and better means of counteracting the poor casting properties of alloys has led to the following casting methods:

1. Crystallization of ingots in molds cooled by running water (immersion method) [9].
2. Crystallization of ingots under high pressure [10].
3. Crystallization of ingots and castings under all-round pressure in autoclaves [17].
4. Crystallization under pressure of a plunger [18].
5. Application of pressure under mold caps and the application of gas pressure in closed risers.
6. Use of atmospheric pressure transmitted by tubes to the shrinkage zone.
7. Methods of continuous casting (and pulling) of ingots and other products.
8. Production of additional pressure by centrifugal force in a rotating mold.
9. Use of vibration of molds filled with metal [20, 22, 23].
10. Acceleration of the solidification of ingots of a weight above 10 tons by the introduction of fusible crystallizers inside the liquid metal in the mold [24].
11. Vacuum treatment of the metal in the liquid state and during solidification [25].
12. Acceleration of the crystallization process in molds cooled by liquid heat carriers, water and air, and external and internal coolers; also the production of rational solidification and feeding of castings by local and overall heating of molds, hot tops, etc.

13. Production of castings by the vacuum crystallization method [21].

14. Production of castings from high-melting metals (molybdenum, titanium, etc.) by build-up arc welding using expendible electrodes.

All these methods are based on improvements in the casting and solidification of the metal and on variations in the pressure above the solidifying metal. Some of them are being successfully employed in foundries for the production of a wide variety of castings from different alloys. An exact knowledge of the pecularities in the behaviour of alloys when solidifying makes it possible to produce high-quality castings and to realize the mechanization and automation of foundry processes, including the production of castings in automated foundries. The applicability of these methods for casting different alloys is far from being uniform, and insufficient study has been given to it.

In regard to questions concerning the dependence of the casting properties of alloys on their composition and the form of the constitutional diagram, it is important to note a striking example of scientific prediction made by D. K. Chernov in the following words: "In all probability, there is a close link between the ability of a crystal to develop correctly and the properties of a metal to pass rapidly from the completely liquid state to the solid state, without a transition of greater or lesser duration through a pasty stage, impeding the rapid and correct development of the crystal growths" [10, p. 175].

Many years of development of the science of metals were required before this hypothesis received its confirmation after large numbers of constitutional diagrams of alloys had been constructed and studied. A prominent part in the study of constitutional diagrams and their connection with the change in the physicochemical, mechanical, and industrial properties of alloys was played by Russian and later by Soviet scientists, headed by Academician N. S. Kurnakov.

For the development of the theory of alloys, and in particular the theory of the casting properties of alloys, the most important significance attaches to the work of A. A. Bochvar and his co-workers on the investigation of the relationship between the variation of these properties and the variation in composition and the form of the constitutional diagrams of alloys. This fruitful line of development in the theory of alloys, based on the physicochemical analysis of N. S. Kurnakov, led to the characteristic diagrams of "composition – casting property (shrinkage, hot-shortness, hydraulic resistance, etc.), which play a considerable part in the theory of alloys and foundry practice. The scientific hypotheses and discoveries of A. A. Bochvar in the field of nonferrous and light alloys and by N. T. Gudtsov in the field of steel, as well as the investigations of their pupils, have made it possible to generalize from an enormous quantity of factual material relating to both old and new methods of producing castings and to reveal the causes of many phenomena and discover their peculiarities when widely varying casting methods are used.

The number of "composition—casting property" diagrams of alloys is insignificant in comparison with the number of diagrams of "composition—service property" (strength, hardness, electrical conductivity, heat-resistance, etc.). According to the assessment made by A. A. Bochvar, in the lack of knowledge of the casting properties of alloys, foundry practice stood at the same level in prewar years as the study of metals before there were any constitutional diagrams. One of the causes of such a situation would appear to be the peculiar complexity of the observed phenomena involved in the process of casting and crystallization of alloys.

In recent years, the theory of the casting properties of metals has undergone some development, but it still lags behind the requirements of industry.

The writer has studied the relationships in the variation of the complex casting properties of nonferrous metals as a function of composition and the form of the phase diagram in the case of binary alloys, and has made an attempt to extend this relationship to alloys of ternary systems. Apparently, for the first time, an attempt has also been made to establish a connection between such properties of alloys as surface tension and fluidity on the basis of a systematic study. Considerable space is devoted in the book to a study of the laws governing the formation of crystallization cracks during the solidification of different alloys in conditions of impeded shrinkage, and to a study of the conditions of the origin of disperse porosity, "black" fracture in magnesium alloys and so forth. Modification and the solidification time of alloys are also dealt with as casting properties.

The author hopes that the theoretical and experimental material he has collected will help to fill in gaps in the present theory of alloys in regard to the relationships governing the variation in casting properties, and to make use of these relationships in solving the general problem of discovering alloys having predetermined properties.

The author assumes that this information will increase the equipment of foundrymen in their efforts to improve the quality of castings, increase productivity and develop foundry production, in accordance with the resolutions of the 21st Congress of the Communist Party of the Soviet Union.

The author pays tribute with deepfelt gratitude to the memory of Academician N. T. Gudtsov, whose valuable advice during the execution of the work and its discussion of the results greatly helped in the completion and publication of some of the chapters. The author considers it his pleasant duty to express his indebtedness to Academician A. A. Bochvar, with whose cooperation and guidance many sections of this book were completed. The author is grateful to colleagues of the Institute of Metallurgy of the Academy of Sciences USSR who participated directly in the completion of various sections of the book: E. S. Kadaner, A. A. Bychkova, V. G. Yudin and also engineers L. N. Kuznetsova, N. P. Dorofeeva, and N. P. Kosorukina, who assisted in the work on magnesium alloys under industrial conditions.

Chapter I

STRUCTURE OF LIQUID METALS AND ALLOYS

Modern Views on the Structure of Pure Liquid Metals

The study of the casting properties of alloys is closely bound up with the study of the structure and properties of liquid metals and with crystallization theories.

Practical data and modern physicochemical theories agree that liquid metals and alloys near the melting point do not possess a disordered, amorphous structure – as appeared formerly to investigators who found a resemblance between the structure and properties of substances in the liquid and gaseous states – but a structure resembling the crystal structure of the solid metal.

The fullest and widest development of the physical theory which sheds light on the nature of liquid substances, solutions and melts, is to be found in the writings of the Soviet and foreign scientists Ya. I. Frenkel', V. I. Danilov, D. Bernal, P. Debye, and others. These theories are important to metallurgical science because they are working hypotheses which explain the mechanism of the melting of metals, the chemical interactions in melts and the mechanism of crystallization, all of which is of great practical importance. Starting from the fact that substances in the solid and liquid states are similar in density and thermodynamic characteristics (for example, specific heat), Ya. I. Frenkel' came to the conclusion that the atomic interactions and interatomic distances should be approximately the same in a solid after melting as before melting. Such a concept of the nature of liquids is shared by leading scientists in Russia and elsewhere.

In his earlier book (1935), V. I. Danilov analyzed the results obtained by various authors in the x-ray study of the structure of water, organic and inorganic liquids and some molten metals. He showed that despite the high mobility of the molecules in liquids, there is a regularity in their arrangement. The ordering of the particles in liquids occurs "as the result of forces similar to those which bind them together in crystals in the solid state. At the same time, probably for all liquids, as the freezing point is approached, the arrangement of the molecules changes, approaching that occurring in crystals of the same substance" [29].

In a paper read at a meeting of the Faraday Society in 1935, D. Bernal stated that the interference of x-rays in liquids is the broadened interference reflection in crystals. In other papers presented at the same meeting, including one by Ya. I. Frenkel', evidence was given to show that liquids near the melting point are no longer completely disordered and that a crystal-like structure predominates in them. In one of the papers presented at this meeting it was shown that the structures of the two metals lead and tin in the liquid state were identical [30].

In his work for the Nobel Prize Award, P. Debye showed that the presence of the interference bands produced in the scattering of x-rays by liquids, even monatomic liquids, points to their quasicrystalline structure, characterized by a definite ordering in the motion and position of the molecules [31].

The work of Professor V. K. Semenchenko on the theory of solutions [32] and the surface tension of metals and alloys (amalgams) [33] shows that the particles in liquids are arranged much more regularly than was formerly supposed. In metals such as mercury and gallium, in the liquid state at temperatures close to the solidification point, the atoms retain an arrangement similar to that which they have in the solid state. The thermal motion of the particles in a molten metal is thus subjected to an interaction which limits their mobility; in the course of a certain period of time, the atoms occupy some mean position in a group of the same atoms. It is this which produces what is called "close-range order" in a liquid, which differs from the "long-range order," that is to say from the strict regularity of the crystal lattice of the solid metal.

In the solid state close to the melting point, the disorder in the coordination of the crystal lattice is insignificant, but it increases abruptly at the melting point, when the lattice breaks down and the distance between the atoms suddenly increases. This transition is accompanied by an abrupt change in volume and electrical conductivity, which for many metals is reduced by more than half. As the temperature increases, the amorphous nature of the liquid gradually increases, but the bonding forces between the atoms continue to be effective; the complete breakdown of coordination occurs only at a considerable distance from the melting point.

The effect of these forces is observed experimentally in the form of the regular x-ray scattering curves obtained by V. I. Danilov and I. V. Radchenko for liquid lead, bismuth, and tin [34] and for the eutectic alloys Pb-Bi, Sn-Bi and others [35]. These curves clearly indicate the orderliness in the arrangement of the atoms in a liquid, similar to that occurring in a crystal.

The results of x-ray analyses of liquid metals are in agreement with the theoretical data for the "blurred" lattices of these metals in the solid state, confirming the correctness of the theory of the quasicrystalline structure of molten metals near the melting point.

In our opinion, such a theory of the structure of pure, liquid metals is confirmed by the practice of single-crystal production. It is obvious that without the ordering of the particles of the liquid metal near the melting point, it would be impossible to pull from the liquid single-crystal specimens attaining several centimetres in diameter. In this case, despite the slow rate of pulling of the single crystal, it is difficult to assume that the multitude of atoms disorderly distributed in the liquid "suddenly" occupy the places prescribed for them in the lattice of the seed crystal. It is evident that this requires some preparatory ordering process of the atoms in the liquid metal.

Structure of Actual Liquid Metals and Alloys

The structure of liquid metals as described above is mainly true for very pure metals. It is natural that the structure of actual metals containing a definite quantity of soluble and insoluble impurities will differ from the structure of an "ideal" metal. This is shown by the fact that, for instance, in one cubic millimeter of practically pure liquid copper containing 0.01% by weight of iron, nickel, and cobalt altogether, the number of impurity atoms is about 10^{17}.

The change in composition of alloys, commencing from small amounts of impurities, will be reflected in both structure and properties of the alloys — viscosity, surface tension, oxidizability, rate of diffusion processes and also on the position of the solubility curves (liquidus and solidus temperatures) and so forth. This change in the liquid properties of alloys acquires special significance when deoxidization, desulfurization, and other metallurgical and foundry processes are carried out, the results of which appreciably affect the properties of the semimanufactured and finished products (for example, segregation of impurities and the formation of crystallization cracks, hot-shortness during rolling, reduction in the plastic properties of alloys at the ordinary temperature and others).

The study of liquid metals, their properties and the connection between the properties of the liquid and solidified metal was considerably developed by the work of Academician A. A. Baikov, who emphasized the importance of the chemical reactions taking place in the liquid metal and the influence of the products of these reactions on the properties of the metal. He ascribed great importance to the condition of the impurities in the liquid metal, either in solution or in "plankton" form (insoluble small liquid or solid slag inclusions, particles of furnace lining, difficultly reducible suspended oxides) [16]. Different impurities behave differently on crystallization and affect the structure and properties of the metal in different ways. Thus, the whole of the "plankton" is trapped by the crystals and is arranged principally in their interior in the form of inclusions of foreign matter. On crystallization, an impurity dissolved in the liquid metal will be situated in that part of the liquid mass in which the crystallization process terminates and which constitutes what is called the "eutectic alloy." Obviously, this structure of the liquid metal bath ("turbid" or "transparent") has a definite influence on the course of crystallization, on the mechanism of crystal formation and growth and on the properties of the metal in the process of solidification. The concept that there is a relationship between the properties of the solid and liquid metal is also developed in the work of Academician N. T. Gudtsov. In his opinion, among the fundamental questions of the study of the properties of the steel ingot, the precrystallization condition of the liquid steel is of considerable importance and ought to be studied alongside a study of the relationship between chemical composition, structure, and properties of different steels in the solid state. N. T. Gudtsov considers that new methods of investigation should be sought, in view of the particular difficulties of studying the properties of liquid steel in the precrystallization period [24].

K. P. Bunin, by centrifugalizing liquid Sn-Bi eutectic, found that it contains portions with a preferred content of one of the components. When specimens 50 mm in length were centrifugalized at a speed of 3000 rpm, opposite ends of the specimens showed a difference in bismuth content equal to approximately 10 at. %. This could be due to the fact that during centrifugalizing, the field of force which developed exceeded that due to the earth's gravity by 2000 times and "projected" the heavier bismuth to the periphery of the rotating mass of the liquid alloy. K. P. Bunin assumes that for this to take place, the mean volume of the portions of bismuth in the liquid eutectic should correspond to 50 unit cells [37]. This investigation confirms that in similar alloys in the precrystallization period, the bond forces between like atoms of the solution begin to predominate over the forces of attraction between unlike atoms, resulting in microheterogeneity of the alloys. K. P. Bunin assumes the possibility of the existence of composition fluctuations in liquids and states that there is a connection between the magnitude and number of such fluctuations: the smaller their magnitude, the greater their number. On the basis of these fluctuations, spontaneous crystallization is possible, apart from that due to crystallization centers in the form of foreign inclusions [38].

Evidently, these groupings of like atoms in liquid alloys are nucleation centers for two forms of crystal preceding the start of the process of eutectic crystallization. Indeed, according to A. A. Bochvar's theory of eutectic crystallization, at the commencement of solidification, crystals of each of the phases are nucleated and grow separately, and it is only after they have come into mutual contact that the relatively rapid process of strictly eutectic crystallization commences [39]. In the opinion of A. A. Bochvar, it is possible that in the precrystallization period, density fluctuations (without the formation of a phase interface) and nuclei, i.e., already existing surfaces of separation, exist simultaneously in the liquid [40, p. 53].

In Prof. E. G. Shvidkovskii's book [41], a picture is given of the structure of a liquid on the basis of experimental data. From data on the variation in viscosity of liquid metals and alloys and x-ray analyses, the author assumes that some of the atoms of a liquid are in the form of ordered groupings in regions of the order of tens of angstroms; the other group of particles of the liquid are moving in complete disorder. Thus, in a liquid, there are motions characteristic of a solid (oscillations about a position of equilibrium) and of a gas (random migration), and any particle can pass from one group to the other. It is natural to assume, says E. G. Shvidkovskii, that immediately after a solid has melted, the liquid consists of ordered quasicrystals and separate migrating particles. The latter are more numerous close to the lattice defects, for instance close to the boundary of separation of the solid phase. In the central regions of such quasicrystals, there is maximum order of atomic arrangement and this diminishes towards the periphery; as the temperature increases, so also does the number of "free" particles and the "erosion" of the ordered regions. Further increase in temperature results in the disappearance of traces of structure and a transition to a state of chaotic distribution of particles and to vapor. However, when the temperature falls, the interatomic forces become more and more evident, bringing the atoms into the crystal lattice, a process which is accompanied by an increase in viscosity.

The foregoing indicates the presence in liquid metals and alloys of groups of atoms and the probability of their existence and their increase in stability as the crystallization point is approached. These concepts have become a permanent part of modern science and have been confirmed by recent work on metallurgy [40, 42] and physics [43]. They enable one to say that, on the change in concentration of alloys, when crystallization "beginnings and ends" are present in them, as determined by the liquidus and solidus temperatures, precrystallization phenomena will assume considerable magnitudes. The structure of liquid alloys, taking shape under the effect of temperature and concentration, will form the basis for the examination of the relationships in the change in physicochemical and technical (casting) properties of liquid metals and alloys.

The concept of the grouping of like atoms in a uniform liquid solution, the partial separation of the latter into layers near the eutectic point is, in our opinion, to some extent in agreement with the separation of liquid metallic alloys into layers of two immiscible liquids. Evidently, these groups are the initial stages of the macroscopic separation of the solutions. In this case, the difference in the individual properties of the components of the solution is so great (for example, the difference in atomic volumes is very great) that the mutual existence of the atoms in a homogeneous solution is impossible, and the uniform liquid solutions separate entirely into two solutions or into the pure components (Al–Pb, Zn–Pb, Cu–Bi and other alloys).

In attempting to correlate the concepts of V. I. Danilov, K. P. Bubin and others on the submicroscopic grouping of atoms of the components in uniform liquid alloys with the critical points in the constitutional diagrams, it must be borne in mind that the majority of eutectics in binary metallic alloys are not composed of the pure components and solid solutions based on them, but of solutions and chemical compounds, or even of several chemical compounds. In such cases, the breakdown of a uniform liquid into two solid varieties may be preceded by more varied and more considerable changes in the structure of the liquid solutions than in the case where the eutectic is formed from the pure components (for example, precrystallization grouping of the atoms in chemical compounds and the like). It may also be assumed that if the lattices of the components of the solution are related and their parameters are close to each other, precrystallization phenomena, consisting in the grouping together of like atoms, will be developed less strongly; in this case, separation of crystals (Ag–Au, Cu–Ni, Ti–Zr, Fe–Cr and other alloys). In the case of the crystallization of a solution, the composition of which, however, corresponds to a pure chemical compound, it is to be expected that there will be an association of atoms in the solution corresponding to the lattice of the cm compound (Al–Sb, Mg–Sn, Mg–Sb and other alloys). In this case, the liquidus curve will show a definite maximum; smoothing of this maximum will be evidence of dissociation of the compound and instability of the chemical complexes in the liquid.

We may also say that "structurization" of the liquid solution close to crystallization temperatures will be increased at concentrations of the solution for which the solubility curve in the liquid state (liquidus curve) is steeper or the maximum is more pronounced. To some extent, this points to a grouping together of atoms from which the higher melting phase is formed, this not infrequently being a chemical compound. This is also in agreement with the well-known circumstance that when a mixture of phases is remelted, the last to melt will be the higher melting crystals, and traces of their existence may be preserved and affect the structure of the solution and the results of its subsequent crystallization.

In conclusion, it must be said that in addition to the x-ray method of studying the structure of liquid metals and alloys, it is also possible to base conclusions regarding their structure on the physicochemical properties of liquids — surface tension, viscosity, electrical conductivity, etc. In their turn, these data may be used in assessing the technical (casting) properties of alloys.

Chapter II

THE CRYSTALLIZATION OF METALS AND THE CONSTITUTIONAL DIAGRAMS OF ALLOYS AS BASIS FOR THE THEORY OF CASTING PROCESSES

Modern Views on the Crystallization Process

An enormous number of investigations have been devoted to a study of the transition of metals and alloys from the liquid to the solid state. This is due to the fact that the conditions under which this process takes place have a considerable influence on the structure of the metal and, consequently, on its technical and service properties.

The crystallization process, like any phase transition, is possible under conditions for which the free energy of the system is lowered. If, at a certain temperature, the solid phase has less energy, the process of increase in the solid phase, i.e., crystallization, will take place; in this case, when at a certain temperature the liquid phase has less energy compared with the energy of the crystalline phase, disappearance of the latter will take place (melting process).

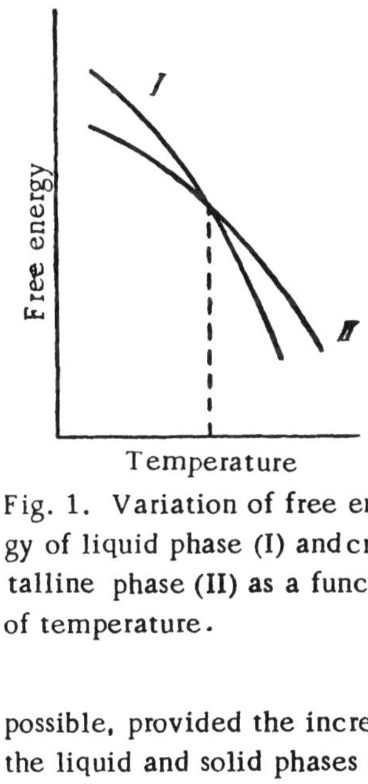

This thermodynamic basis of the processes of melting and solidification is usually represented graphically by curves of the dependence of the free energy of the phases on temperature (Fig. 1). The point of intersection of the free energy curves corresponds to equilibrium between the crystalline and liquid phases: the passage of atoms from one phase to the other does not alter the energy of the system; a shift toward a lower temperature makes the solid phase more stable. Thus, this point defines the melting or crystallizing temperature of the substance. Since the crystallization process can take place only with lowering of the temperature relative to the equilibrium temperature, under actual conditions, some supercooling of the liquid phase is essential for crystallization; at the same time, the liquid phase will be less stable than the solid phase and phase transformation will be possible.

Fig. 1. Variation of free energy of liquid phase (I) and crystalline phase (II) as a function of temperature.

The second condition for continuation of the initial crystallization stage is that the crystal nuclei, formed in the liquid at the given temperature (with given supercooling) should be stable.

Such stability of the nucleus (in the thermodynamic respect) is provided when the nucleus assumes a certain "critical" size or exceeds that size. This is possible, provided the increase in free energy due to the appearance of a surface of separation (interface) between the liquid and solid phases is compensated "by its reduction, due to the transition of some volume of the less stable phase into a corresponding volume of the stable phase" [40, pp. 52, 53]. Such compensation becomes possible with increase in dimensions of the nucleus, since the reduction in the free energy in the volume is proportional to the cube of the radius of the nucleus, while the surface energy increases in proportion only to the square of that radius.

Thus, the instant the nucleus attains the critical size, the free energy of the system begins to decrease, the nuclei become stable and can form crystallization centers. This explains the influence exercised on crystallization processes by precrystallization phenomena in the liquid, the presence in the latter of quasicrystalline regions, the structure of which may correspond to the structure of the forming crystals. Such concepts of the crystallization processes of metal are widely held at the present time (A. I. Shubnikov, A. A. Bochvar, V. I. Danilov, Ya. S. Umanskii and others). As we shall see, in these concepts, questions of crystallization theory are closely bound up with concepts of the structure of liquid substances and the precrystallization phenomena in a liquid.

The fact that the liquid metal attains a metastable state at a definite, even though slight, degree of supercooling plays a decisive part in the transition to the solid state. The capacity for supercooling obviously depends mainly on the internal structure and energy properties of the atoms of the metal. According to V. I. Danilov, bismuth, tin, and mercury supercool most readily and to the greatest extent, that is to say, metals in which the crystal lattice is comparatively complicated and not very compact [44]. In the crystallization of aluminum, it has not been possible to discover any appreciable supercooling irrespective of the degree of superheating of the liquid metal. Evidently, in this case, the nucleation of stable crystallization centers is facilitated by the similarity between the arrangement of the atoms in the liquid in the precrystallization period and their arrangement in the crystal lattice, and consequently by the low energy of formation of these centers.

It is customary to consider the commencement of the crystallization of liquid metals (just like other substances) to be the formation of stable ("critical") crystallization centers and also the subsequent growth of these crystals (or "embryos," as D. K. Chernov called them).

These two fundamental conditions play a decisive part in the theory of phase transformations and determine the result of crystallization. If the degree of supercooling is very slight or if supercooling is absent altogether, both crystallization parameters (number of centers and rate of growth) are of very little significance. They become much more important as supercooling increases up to a certain limit, and then their importance decreases, since the atoms lose their mobility in a strongly supercooled liquid.

At a definite degree of supercooling, some liquids (salol and the like) can be kept in the form of an amorphous substance for a long time, and a seed must be introduced into them to bring about their crystallization. It is impossible to prevent the crystallization of liquid metals, due to the low energy of formation of equilibrium nuclei, together with the above-mentioned quasicrystalline structure of liquid metals near the melting point.

Figure 2 shows the dependence of the number of centers and the linear rate of growth on temperature (supercooling). The maximum linear rate of crystallization is reached at a lower degree of supercooling than is the maximum number of crystallization centers.

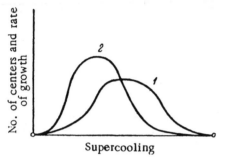

Fig. 2. Dependence of number of centers (1) and linear rate of growth of crystals (2) on temperature.

According to a number of authors [42], these curves which were first constructed by G. Tammann on the basis of observations of the crystallization of organic substances in small volumes, retain their significance in the modern theory of the crystallization of metals. A. A. Bochvar considers the curves of the linear rate of transformation to be of considerable interest to metallurgists, since they provide a qualitative picture of the process and indicate practical ways of acting on it. In his opinion, "the principal method of the nucleation of crystals in a liquid is their nucleation on the surface of impurity inclusions or on external interfaces." The structure of an actual ingot is, therefore, mainly determined by both the content and character of the impurities [40].

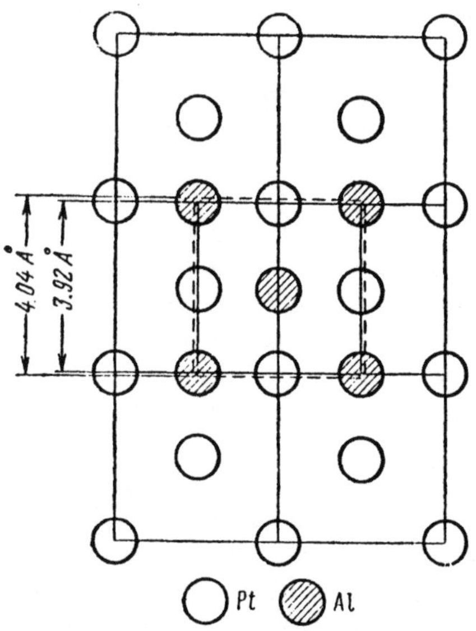

The kinetics of the transition of a liquid metal to the solid state thus depend in a decisive manner on the presence in the liquid of existing interfaces in the form of insoluble impurities (A. A. Baikov called them "plankton" in liquid steel), which are always present in actual liquid metals and alloys. Such foreign nuclei facilitate the initial crystallization process, since they assist in reducing the free energy, whereas growth to a critical size of a nucleus produced from the atoms of the metal itself is accompanied, according to Gibb's theory, by an increase in the free energy of the system. This initial increase in energy is produced by the work expended in creating the interface between the nucleus of the crystal and the liquid.

The energy required for the production of a stable nucleus and for its growth will be a minimum when the configuration of the atoms of the nucleating crystal are similar to the arrangement of the atoms in the "support," i.e., the impurity, that is to say, when the principle of the orientational and dimensional correspondence of the phases participating in the process, as formulated by P. D. Dankov [45, 46], is clearly exhibited. According to this principle, the formation of crystallization centers is due not so much to an accidental arrange-

Fig. 3. Diagram of the geometrical complex of the growth of aluminum on the surface of a cube face of platinum (the outline of the normal aluminum cell is shown in dash lines).

○ Pt ◍ Al

4.04 Å 3.92 Å

ment of the atoms, resembling the lattice of the crystallizing metal, as to existing interfaces, practically always present in actual metals. An important part is played in this by the similarity of the lattices and the nearness of the atomic distances of the support and of the nucleating crystal (Fig. 3).

The experimental data of V. I. Danilov showed that the presence of oxides of low-melting metals facilitates the crystallization process of such metals, since the degree of supercooling is reduced. Thus, in experiments with pure bismuth, without oxides, supercooling was 35 - 45°C, and in the presence of oxides it was only 10 - 12°C; for lead, these figures were 7 - 8°C and 3 - 4°C, respectively [47]. In another investigation, V. I. Danilov found that the nucleation of crystallization centers of potassium occurs only on oxides [48]. These, as well as previous investigations of V. I. Danilov, led him to conclude that, at the present time, "it may be said with certainty that, in liquids, under normal cooling conditions, crystallization centers are not produced spontaneously, but on impurities" [49].

In the crystallization of antimony and bismuth, the phenomenon of supercooling can be observed quite easily. Evidently, in this case, the probability of the formation of intrinsic centers is relatively slight, but the redistribution of the atoms in the liquid and their "arrestment" in a position corresponding to the lattice of the solid metal, at temperatures near the melting point, occurs with much greater difficulty than in the case of other metals. This may be due to the fact that the atomic volume of these metals increases on solidification, and does not decrease like that of most metals.

Summing up modern views on the crystallization processes of metals, it must be emphasized once more than in perfectly "transparent" liquid metals, the crystallization nuclei are produced as a result of the self-diffusion of groups of atoms of the metals. It may therefore be expected that there is a connection between the tendency of metals to supercooling and their properties, such as variation in density, surface tension, viscosity and the like. In actual metals and alloys, the atoms of extraneous impurities and nonmetallic inclusions exercise a decisive influence on the crystallization process.

Crystallization Processes and the Structure of Metals and Alloys

The crystallization process and its results may be observed directly by using small volumes of a solidifying transparent substance, similar to that described in the work of Academicians A. V. Shubnikov [50] and A. A. Bochvar [39], and also in the work of other investigators, one of whom made a cinematographic study of this process [51].

The most characteristic of this work is the examination of the form of crystals and crystalline formations and its variation under the influence of temperature (crystallization rate), the composition of the melts and other factors. The universal form of the crystal is recognized to be the dendrite, and the study of its nature, kinetic growth, and formation has been the subject of much work by outstanding scientists, commencing with A. K. Chernov [10, 11, 16, 27, 39, 42, 44, 45, 50 and others].

An international questionnaire on the dendrite, conducted in 1932 by the journal "Metallurg," in which Academicians A. A. Baikov, N. T. Gudtsov, and A. A. Bochvar, together with a number of foreign scientists took part, showed the complexity of this problem and the different views on it held by individual scientists [52]. The questionnaire resulted in the instigation of a number of investigations and left a deep impression on the development of crystallization theory and the science of metals.

Currently, due to a number of investigations published in the form of monographs and collective works [39, 50 - 58, 60], this problem is gradually being solved. It is now possible to say more definitely that a dendrite is crystal growing from one center. A main crystalline axis and axes or branches of second and higher orders are distinguished in it. The directions of the principal crystalline axis of a dendrite and its side branches are the directions of growth, which for the production of castings in molds are opposite to the direction of heat removal. The magnitude of a dendrite, that is to say, its volume and the extent along the individual growth directions (in a given volume of liquid), may be regarded as the result of the competitive "survival" of individual dendrites and the suppression of the growth of their neighbors, due to the change in direction of the crystallization front or individual portions of it under the action of fluctuating thermal fields or waves, and also due to a considerable amount of overtaking in the growth of the side branches (the latter may give rise to the formation of independent dendrites growing in a different direction). An important part in this is played both by the quantity of substance solidifying in the given volume, that is to say the quantity of heat of solidification liberated, and the coefficient of thermal conductivity of the metal, which varies with the temperature and orientation of the growing dendrites. It is also not unlikely that individual dendrites are the result of the recrystallization interpenetration of several dendrites or their parts, having similar directions of crystal lattice.

These considerations regarding the causes of formation of dendritic crystallization may be related mainly to the crystallization of pure (ideal) liquids, not containing foreign atoms and molecules, the accumulation of which on the crystal–liquid boundary may vary the composition, crystallization temperature, and thermal conditions of solidification of the given volume. Molten metals are actually far from being ideal liquids, and the presence in them of foreign impurities, suspensions and the like cannot by any means be ignored. Therefore, as pointed out by A. A. Bochvar [39, 40], in examining the causes of dendritic crystallization, it is necessary to bear in mind the "natural" obstacles to growth in the form of accumulations of impurities and the variations in composition in the crystal-liquid boundary zone. The scale of the variation in composition may be judged from the example given later (see p. 17). As the result of overcoming such obstacles to growth, the dendrite becomes more branching, its side branches tend to "penetrate" the portions of the liquid in which the boundary zones are not yet separated from the growing crystal by a barrier consisting of a layer of foreign particles. If diffusion and convection, which reduce the concentration of impurities in the boundary zones, are only slightly developed, the growth of the dendrites is reduced and the degree of their branching becomes very considerable; the structure of the solidifying metal will show clear signs of segregation, both between the dendrites and within them. Segregation problems have been explained in detail in a recent book by I. I. Golikov [235].

Leaving aside the peculiarities of crystallization in the presence of small quantities of surface-active impurities, or impurities entering into definite chemical reaction with the forming crystals, it should be pointed out that in examining the question of crystallization and the mechanism of the formation and growth of crystals, it is important to bear in mind the profound difference between the crystallization processes of very pure and pure <u>metals</u> or chemical compounds, and <u>alloys</u>, which solidify in a temperature range, when the composition of the solid and liquid phases, or at least of only one liquid phase, varies during the crystallization process. In the first case, the transition of the metal to the solid state occurs generally speaking at constant temperature, without the formation of layers of foreign atoms at the crystal-liquid interface, and the possibility of linear growth of dendrites is comparatively great; in such a case, single crystals can be grown principally by making use of the cooling conditions on the seed (growing single crystal) –liquid metal interface. In the crystallization of industrial metals and alloys, the size of the dendrites, the degree of their branching, and the magnitude of the dendritic cells depend both on the cooling conditions of crystallization and on the rate of diffusion of the atoms in the solution (melt) and the convection currents in the latter. In this case, the higher the rate of cooling, i.e., the more strongly obstructed is the movement and influx from the solution of atoms of the substance forming the dendrite, the greater the possibility of the occurrence and growth of fresh dendrites, both in the melt and on the side branches of dendrites and, in general, on any already existing surfaces.

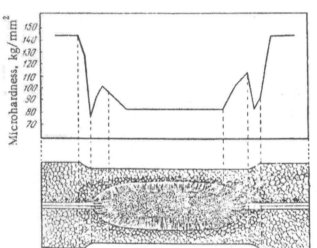

Fig. 4. Formation of crystallization zones in a spot weld on duralumin sheets.

A study of the crystallization process of steel ingots, a process through which an enormous mass of industrial metal "passes", led N. T. Gudtsov to the theory of the discontinuous crystallization of ingots, previously foretold by D. K. Chernov. A considerable amount of work goes to show that this theory is in agreement with practice, since actual ingots show a periodicity of the various mechanical properties, chemical composition, and content of non-metallic impurities corresponding to periodicity in crystallization [11, 53, 55].

It should be pointed out that the tendency to form zones of columnar and equiaxial crystals may occur in both large ingots and very small volumes of crystallizing metal. The formation of such zones is often observed in welds of different thicknesses, well indicating the character of crystallization directed away from the heat-dissipating front. Figure 4 shows as a characteristic example, the structure of duralumin at the position of a spot weld of two rolled sheets, each 2 mm thick. The peripheral zone of the molten body of metal has solidified in the form of a regular rim of rodlike crystals – dendrites, with a length of about 0.5 mm, forming from $1/4$ to $2/3$ of the total thickness of the fused zone. The central zone of the melt has crystallized in the form of equiaxial dendrites having a diameter approximately equal to the diameter (thickness) of the rodlike crystals; the actual diameter of the latter is about 0.05 mm.

In the practice of the production of castings of alloys of different composition and weight, there is an infinite variety in the size of the dendrites themselves and in their internal structure. Thus, D. K. Chernov's dendrite, familiar to all metallurgists, which was extracted from a steel block, was 39 cm long, while in the production of ingots from duralumin by the continuous casting process [57], and of magnesium alloys by the immersion method [236], separate threadlike crystals having a length of several meters can be seen in their structure, this being the direct consequence of directional crystallization and continuous growing of a dendrite into the liquid melt [57]. Figure 5 shows the junctions between dendrites with different growth directions, Fig. 6, the internal structure of aluminum dendrites,

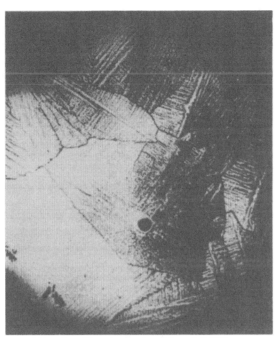

Fig. 5. Aluminum dendrites (× 16).

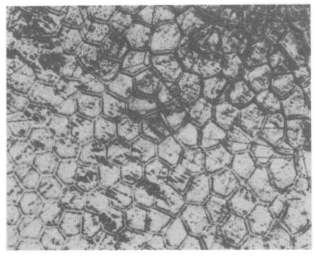

Fig. 6. Internal structure of dendrites of pure aluminum (× 30).

and Fig. 7, the character and scale of the variation of dendritic cells in duralumin sheets under the influence of the crystallization rate [56]. Undoubted interest is to be seen in the fact that with the enormous decrease in the size of the dendrite branches (dendritic cells) on increase in the rate of solidification (reduction in the cells to one-thirtieth or one-fourtieth with increase in the rate of pulling −solidification from 0.3 - 0.8 to 60 - 80 cm/min), the size of the dendrites themselves does not alter appreciably. A similar picture of the variation in structure of steel under the influence of the rate of cooling has also been described by B. B. Gulyaev (Fig. 8).

The limit case with regard to the size of the crystallite (dendrite) and its internal cells (branches) will obviously be a single crystal of the purest substance, pulled slowly from the melt, every precaution being taken to avoid any disturbance in the regular course of the gradual build-up of atomic layers on the already existing crystal surface (cooling of the seed, stirring of the melt, rate of pulling the crystal, etc.). Work in this direction has been developed recently in connection with the production of the semiconductors germanium and silicon [60].

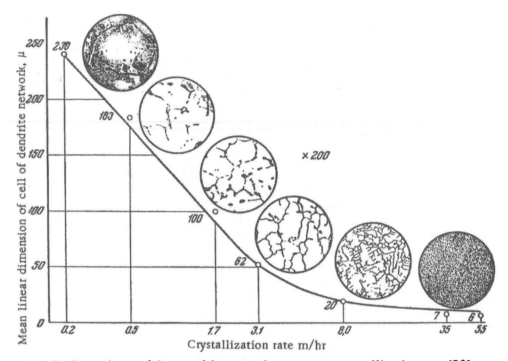

Fig. 7. Dependence of degree of fineness of structure on crystallization rate [56].

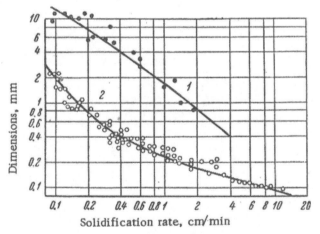

Fig. 8. Influence of solidification rate on the size of dendrites [53, 55]: 1) Cast in sand; 2) cast in metal molds.

The practical results of the crystallization process of metals and alloys observed on fractures, plates, and polished surfaces make it possible to a certain extent to retrace the course of solidification of castings and ingots and assess the influence exerted on it by various metallurgical factors, methods and conditions of melting and casting, solidification conditions, composition of the alloys and so forth. In addition to this, by examination of polished surfaces, it is possible to ascertain the part played by introduced crystallization nuclei or modifiers, the character of the liberation of gas and shrinkage phenomena in the solidification and further cooling of castings.

It is, however, not possible in every case by comparing the structure and properties of the solidified metal to reproduce the complete picture of the crystallization process of a given alloy and to secure sufficient reproducibility of the results of this process. A striking example in this connection is the discussion carried on in recent years concerning "shower crystallization," and the nature of the formation in steel ingots of the peculiar region, the lower cone, with its "own" structure and special properties, differing from the properties of the metal taken from other, nearby, zones of the ingot.

The existence of this region in ingots was discovered by D. K. Chernov, who gave an explanation of its occurrence in accordance with the then existing state of scientific knowledge of the solidification process of steel. Notwithstanding the fact that, since that time, the steel ingot has been the subject of an enormous number of investigations, a more or less unambiguous solution of this problem has not yet been found.

Thus, the adherents of the "shower" theory of the crystallization of the lower cone [61 - 63] point out that the high purity of the metal in the lower cone is due to the descent of relatively heavy isolated crystals, formed in the absence of segregation impurities [61]. Other investigators have found that the cone contains a large quantity of nonmetallic inclusions entrained by the descending crystals [62]. These crystals may be formed in the stream of metal and in the runners, and also when they are washed away and detached from the solidified ingot wall, and also close to the crystallization front, due to supercooling of the liquid metal [63].

In our opinion the question of the precipitation ("shower") crystallization of steel has been approached in the most correct way by I. N. Golikov and P. V. Kozlov, who pointed out that the process of the formation and development of descending crystals differs in different steels; it is also most frequent in steels having a wide crystallization range [62]. Such a physicochemical basis of the process and its relation to constitutional diagrams are quite logical and, as shown in [61], are confirmed by practical experience in ingot production. The mechanism of the "shower" crystallization of alloys having a wide crystallization range is not contradictory to the views held by D. K. Chernov, N. T. Gudtsov and other investigators on the solidification of a steel ingot, in which the process includes a phase of consecutive crystallization and volume crystallization.

The opponents of the "shower" crystallization hypothesis [64 - 66] do not consider it possible that freely falling crystals, produced in the metal stream and on the free exposed surface of the ingot, can be preserved, since they must be remelted on passing through the thickness of liquid metal at the liquidus temperature.

The possibility of the formation of isolated crystals in the liquid close to the crystallization front, beyond the molten layer of high impurity content, is also disputed. This would be possible only in the case of marked supercooling of the liquid, which no one has really succeeded in confirming, not even with the use of sensitive, inertia-free galvanometers [64]. Other investigators also point out that supercooling on the crystallization front is impossible because of the constant movement of the metal in the mold during solidification [65]. In the solidification of an ingot weighing 6 tons, N. E. Skorokhodov found that the temperature of the liquid steel in the mold was constant for 86 min after pouring and that no supercooling occurred during this period. He also refutes the possibility of individual crystals becoming detached from the walls and surface of the ingot and the possibility of "shower" crystallization being due to this [66].

It should be pointed out that recently this question has again been raised in connection with certain peculiarities in the solidification of duralumin by continuous casting [67, 68]. Accurate temperature measurements close to the crystallization front (bare thermocouple) showed that the liquid alloy was supercooled by 3 - 5°; the formation of free crystals (flakes) was also shown, these being detached from the surface and giving rise to a commencement of precipitation crystallization, the result of which was the formation of accumulations of such crystals ("bright spots") in the castings [68].

The conflicting points of view indicate the inadequacy of our knowledge and methods of studying structure in providing a true picture of the crystallization of alloys and revealing the connection between the structure of alloys of different compositions and the various factors involved. They point to the need for employing in this case new methods of investigation, one of which is to construct composition – property diagrams (in particular "composition–casting property" diagrams), the application of which in many fields of chemistry and metallurgy is giving perfectly satisfactory results [27, 70].

This brief review of the views on the crystallization of metals would be incomplete if we did not indicate the possibility of varying the course of crystallization of a mass of metal by means of vibratory action. Thus, the effect of ultrasonic vibrations in the formation of the structure of tin, zinc, and cadmium has been found to give a considerable refinement of the crystals, but not the elimination of columnar crystallization. Ultrasonic treatment of alloys of tin with zinc produces a sharp change in the structure of the primary crystals [69].

In what follows, the processes of crystallization and structure-formation of pure metals are considered in relation to the presence of impurities in the metals, and the crystallization and structure of alloys are considered in relation to the position of the latter in the constitutional diagrams.

The work of A. A. Bochvar is an important contribution to the theory of alloys and in particular the theory of the crystallization of alloys. Even in the first of his publications on the study of the structure of binary and ternary eutectic alloys, A. A. Bochvar defined the peculiarities in the mechanism of eutectic crystallization in metal alloys, connected with the formation of rims round the primary precipitates in these alloys, followed by the nucleation and growth of the individual components of the eutectic, as the concentration of the component crystallizing out later increases in the liquid. It was found that the possibility of the simultaneous crystallization of the components of the eutectic occurs only toward the end of the crystallization of a certain volume, when the latter contains crystallization centers of both components of the eutectic [70]. In his later work, A. A. Bochvar made a detailed study of the crystallization mechanism of a eutectic and the kinetics of crystallization of the primary and secondary formations in alloys of this type [39]. He found that supercooling and the presence of both components of the eutectic are essential for the commencement of eutectic crystallization; the moment the components come into contact with each other, eutectic crystallization actually commences and proceeds at a relatively high rate − higher than that of the crystallization of the individual eutectic components. This is due to the interface formed at the contact of the two crystals of different phases, on either side of which are collected the dissimilar atoms. At the same time, of course, at the liquid-crystal interface, there is no appreciable enrichment of the liquid in atoms of either of the components and crystallization occurs rapidly.

A study of the crystallization kinetics of alloys as a function of composition showed that the linear rate of crystallization of the primary formations for low concentration of the second component is higher than the rate of crystallization of the eutectic, but near the eutectic point, these rates are close to each other. The maximum of the curves of linear rate of crystallization for a eutectic is always much lower than for the pure components or primary formations in alloys of low concentrations. The relation between these rates for melts of different composition is shown in Figs. 9 and 10.

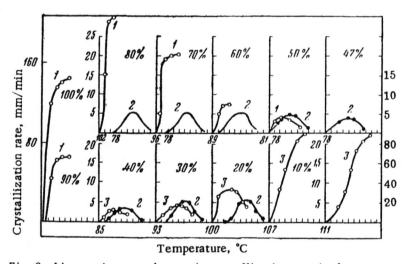

Fig. 9. Linear primary and eutectic crystallization rate in the system acetanilide − dinitrophenol [39]: 1) Primary acetanilide; 2) eutectic; 3) primary dinitrophenol.

The investigation of a number of metallic systems furthermore enabled A. A. Bochvar to appreciate another factor in crystallization − the probability of the nucleation of primary crystals as a function of the composition and the position of the eutectic point in the constitutional diagram. For the case of an ideal equilibrium system, the probability of the nucleation of both kinds of crystals is the same at the eutectic point only; at the same time, the centers of crystals of B produced for example in the field of primary crystallization of component A ought, according to the phase rule, to be dissolved. In actual systems, however, it is not infrequently found that there is a departure from such a law in the case of slow cooling (Fig. 11).

_____

* The term "solidification" is used more often in these cases when the intention is to emphasize the thermophysical aspect of the crystallization process as proposed by B. B. Gulyaev [53].

In our own investigations also, it has been found repeatedly that in the structure of hypereutectic alloys, for example an alloy of aluminum with 15% Si, primary formations of silicon and aluminum are present simultaneously (Fig. 11b). Evidently, the primary crystallization of silicon produces portions high in aluminum, in which aluminum dendrites are formed up to the commencement of eutectic crystallization.

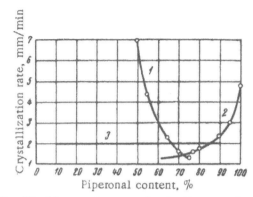

Fig. 10. Linear primary and eutectic crystallization rate at 17°C in the system azobenzene – piperonal [39]: Crystallization: 1) azobenzene; 2) piperonal; 3) eutectic.

As one of the conclusions, the supposition was made that the capacity of a eutectic for independent crystallization is a minimum and the tendency to supercooling is a maximum. The existence of supercooling of a eutectic in the system Al–Si was subsequently confirmed by G. M. Kuznetsov [127].

The position of alloys in the constitutional diagrams of eutectic type thus determines the course of the crystallization process in regard to the probability of nucleation and the magnitude of the linear rate of growth of primary and secondary formations, the relationship of primary and secondary crystallization in the solidification range, the relative quantities of heat of primary and secondary crystallization, and, consequently, also the solidification time of a given alloy. The latter is reflected in the work of A. A. Bochvar and V. V. Kuzina [74] showing that for the same thermal conditions, the time for complete crystallization of the same quantity of pure aluminum was 1.6 to 1.8 times less than for aluminum alloyed with 5 and 12% copper, which is connected directly to the solidification interval, equal to 100 - 80°C for these alloys. Similar relationships of the solidification time (rate) for alloys of a different composition are also given in later work of English investigators [75, 76] and in an article by B. B. Gulyaev and O. N. Magnitskii [237].

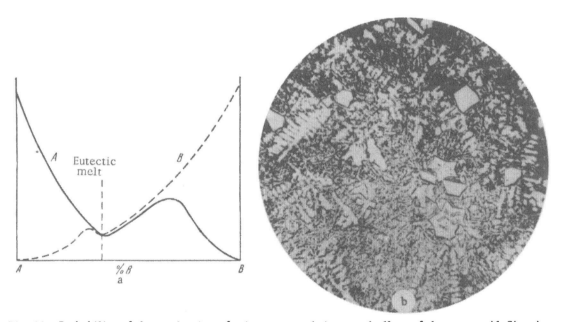

Fig. 11. Probability of the nucleation of primary crystals in actual alloys of the system Al–Si: a) Diagram [71]; b) Microstructure of the alloy Al+15% Si.

The literature contains little published work (apart from that mentioned above) to show the quantitative connection between the form of the constitutional diagrams and the results of crystallization. For peritectic crystallization, there are available the systematic observations of Japanese investigators, showing the part played by primary (relatively higher-melting) formations on the probability of crystal nucleation and on the final result of crystallization [77]. These investigations indicated that if in the crystallization of a given alloy, peritectic crystallization occurs, the size of the crystals diminishes abruptly (Fig. 12). The jump in the change in size of the grain corresponds to the concentration at the transition point indicating the composition of the liquid phase participating in the peritectic transformation. The primary, high-melting crystals separated from melts, the composition of which is somewhat to the right of the transition point, marked by the arrow, do not disappear completely as a result of the peritectic reaction (as is prescribed by the equilibrium constitutional diagram), but remain in the melt as crystallization centers. As a result, the size of the crystals in these alloys is reduced abruptly by a factor of several tens, and in the case of alloys of the Zn–Cu and Zn–Ag systems by a factor of one hundred or more.

For alloys of the solid solution type, spontaneous refinement of the crystals with increase in alloying should also be expected, since other conditions being the same, one of the factors — linear crystallization rate — will be diminished in accordance with the more difficult diffusion and the increasing microsegregation. In this, an important part will be played by the magnitude of the crystallization range. An approximate picture of the formation of two different forms of crystal for solid solutions on a copper basis, solidifying in a small and large range, is illustrated by the diagrams in Fig. 13, from which it follows that the greater the solidification range of the Cu–Sn alloy, the greater is the region of the casting in which equiaxial, relatively small crystals are developed. Cu–Al alloys, solidifying in a narrow temperature range give castings with a clearly pronounced columnar structure.

A good illustration of the relationship between the results of crystallization and constitutional diagrams (or individual portions of them) is provided by the results, published in 1955, of work by M. V. Mal'tsev on the structure and properties of light alloys and nonferrous alloys [78]. He also showed the enormous influence exerted by precrystallization phenomena in the liquid, and the formation of high-melting nuclei in the liquid on the crystallization process and the final structure of castings. Figure 14 shows the most characteristic results of this work. These results require no comment. It must be mentioned that the conclusions of M. V. Mal'tsev to explain the effect of modifiers on the structure of alloys from the point of view of the phase diagrams conflict with the views of some investigators, whose opinion is that the refinement of the crystals (for example in the crystallization of austenitic steel) is due to the effect of surface-active additions of calcium and boron [79]. Evidently, much research work is yet to be done to show in what cases an adsorption mechanism and in what cases a nucleation mechanism of modification occurs in the primary or secondary crystallization of alloys. The results of some of our work in this direction are given in the next chapter.

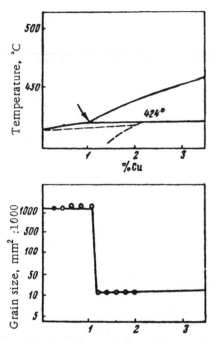

Fig. 12. Influence of the critical points of the constitutional diagram on crystal size in the solidification of alloys of the Zn–Cu system [77].

Thus, the results of the crystallization process of alloys in regard to the final structure of castings are directly related to the form of the constitutional diagram and, in individual cases, are predictable. The variation in the thermal conditions of solidification, the presence of nuclei, surface-active additions and the like in the original liquid may distort this relationship and lead to nonreproducibility of research results. Therefore, in addition to investigations of the structure of the solidifying metal, a study is also made of the relationships expressed by the "composition – property", "composition – temperature – property" and other diagrams, which throw light on the course of crystallization from the liquid, phase-recrystallization and the like. By determining, by different methods, the character of the "composition – property" curves and comparing them with the structure and, finally the form of the constitutional diagrams of alloys, the mechanism and kinetics of the crystallization process, and the influence of different factors on them may be indicated objectively; at the same time it will be possible to ascertain what is typical and what is general in this complex process.

In our work on the study of the casting properties of alloys we have made extensive use of the method and experimental construction of "composition – property" curves and of the analysis of the structure of alloys in relation to their position in the constitutional diagrams.

Constitutional Diagrams and the Casting Properties of Alloys

Investigations of the kinetics and mechanism of the crystallization of pure metals, the primary and secondary crystallization of alloys of various concentrations, and the character of the resulting structures formed the basis of work on the study of the casting properties of alloys and the improvement of methods of producing castings. The

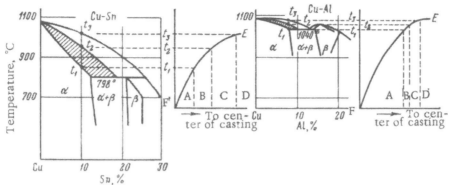

Fig. 13. Influence of the crystallization range on the width of liquid-solid zone in solidification. A) Crystals; B) crystals + liquid; C) liquid + crystals; D) liquid; FE) curve of temperature distribution in casting.

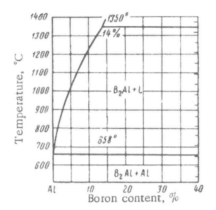

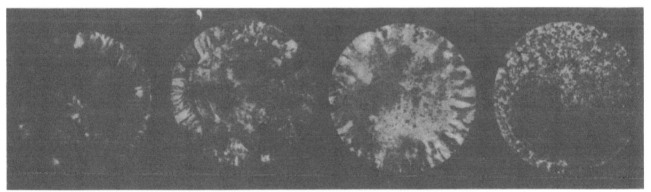

Fig. 14. Al-B constitutional diagram and the modifying effect of boron on Al-Mn alloys: 1) Initial alloy AMts; 2) AMts + 0.05% B; 3) AMts + 0.1% B; 4) AMts + 0.2% B.

most fruitful of these investigations has been the search for systematic variations in the casting properties of alloys with variation in their composition and in relation to the form of the constitutional diagram, since the solidification process of a given alloy and its final structure in the cast state, other conditions being the same, are determined by the position of the given alloy in the constitutional diagram.

The constitutional diagrams of alloys, forming the basis of metallography, predict methods of solving the problem of the choice of alloys of predetermined casting properties, i.e., having a given relative shrinkage, fluidity, susceptibility to hot-shortness, shrinkage porosity and so forth.

Such a method of investigation to ascertain the systematic variations in the casting properties of alloys was used extensively for the first time in the work of A. A. Bochvar and his collaborators. The basis of this work was the experimental construction of "composition − casting property" curves, disclosing a systematic connection between the character of these curves and the process of solidification of alloys as determined by their position in the constitutional diagrams. Curves of a similar kind have been and are being studied in considerable numbers to discover the relationships in the variation of the service properties of alloys. By far the majority of such investigations deal with the construction of "composition – strength," "composition – plasticity," "composition – temperature – strength" and similar curves.

Relatively few such diagrams have been constructed for the casting properties of alloys, and, therefore, the theoretical basis of foundry practice is still on a low level. Currently, the constitutional diagrams of alloys are used as basis in explaining the behavior of alloys in the production of castings and for analyzing the processes of solidification, structure formation, etc.,only in the best Soviet handbooks on metallurgy [40] and foundry practice [7, 8, 80]. In handbooks on the casting of light alloys [81] and steel [82], recently published abroad, constitutional diagrams are not as a rule used for indicating the relationships in the variation of casting properties. It is evident that the number of such investigations has not yet been sufficient to provide a series of "composition – casting property" diagrams to establish corresponding sections of "foundry metallurgy." Such diagrams would undoubtedly facilitate the study of problems concerning the extensive variety of modern methods of casting ferrous, nonferrous, and light alloys (continuous casting and ingot casting, casting in sand, metal and other molds, centrifugal casting).

The emphasis on the important part played by "composition – casting property" curves must not be allowed to dominate the great importance of temperature-rate factors, which considerably modify the conditions of the solidification process and its result. When the technical properties of foundry alloys are being assessed, preference should not be given to any one property in particular, but the combination of several properties must be kept in view. For instance, of the many requirements which foundry alloys must satisfy, that of low shrinkage is given prominence, because it is said to determine the susceptibility of an alloy to hot-shortness; in reality, alloys of a given composition and showing either low or high shrinkage may exhibit considerable hot-shortness or produce castings without cracks.

In his work on the casting properties of alloys, A. A. Bochvar points out that the important part in the determination of a number of these properties is played by crystallization processes, especially crystallization in a temperature range resulting in the properties being directly related to the form of the constitutional diagram. Thus, in work on the study of the casting properties of alloys, which is one of the most important in its practical results, A. A. Bochvar [84] found the systematic character of curves of "composition – volume of shrinkage pores and cavities" applicable to the constitutional diagrams of alloys of eutectic type. Figure 15 shows these diagrams for the case of crystallization in a vacuum and under pressure. The curve AB determining the possible maximum volume of shrinkage voids shows only their total volume but not the character and form of their distribution in the metal. For pure metals and eutectics, the volume variations on solidification will appear mainly as a concentrated shrinkage cavity or pipe, situated in the hottest point of the casting. In alloys solidifying in a temperature range, the volume of the concentrated shrinkage cavity will be less, due to the formation of disperse porosity, the volume of which, for alloys having maximum solidification range, will be equal to the volume of the concentrated shrinkage cavity (for crystallization in a vacuum and under pressure). During solidification under ordinary conditions (under atmospheric pressure), there will be some impregnation of disperse porosity (inter- and intradendritic) and for this reason there will be external shrinkage, as well as a concentrated shrinkage porosity in those alloys in which it is absent when they solidify in a vacuum.

Increase in the pressure prevailing above crystallizing castings will increase the external shrinkage and reduce the volume of disperse porosity in alloys which are in a liquid solid state for a lengthy time. Thus, the "composition-property" curves show that the application of pressure is most effective in eliminating porosity and increasing imperviousness (density) principally for castings of alloys solidifying in a wide temperature range. The application of

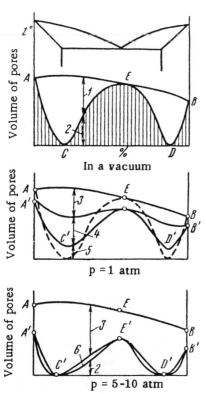

Fig. 15. Dependence of casting properties of alloys on composition and the form of the constitutional diagram: 1) Volume of disperse porosity; 2) volume of pipes; 3) external shrinkage; 4) fine porosity; 5) shrinkage pipe; 6) volume of fine porosity.

the method of crystallization under pressure for the production of castings from pure metals or alloys having a narrow solidification range is not always rational.

Subsequent development of work on the theory of the casting properties of alloys concerned the study of the combination of these properties — linear shrinkage, hot-shortness, solidification stresses, and soundness. This work will be discussed in the corresponding chapters of the book. At this point, we shall merely indicate the chronological sequence of the work. Thus, in 1940 and 1941, work by A. A. Bochvar was published on the theory of linear shrinkage, the investigation of hot-shortness in alloys, on the variation in imperviousness with variation in composition and the form of the constitutional diagrams and so forth. In 1942 and 1943, A. A. Bochvar published work concerning the laws governing the variation of these properties, based on the character of the solidification of alloys and their position in the constitutional diagrams.

In recent years, work in this direction has been developed further and has led to the discovery of new relationships in the variation of the casting properties of alloys with variation in composition and external factors. Reference will be made to investigations connected with a detailed study of the region of constitutional diagrams of binary systems, in which solid and liquid phases exist together, and which previously had been comparatively little studied. These investigations carried out in the USSR and abroad have shown that the properties of the liquid-solid mixture differ sharply in different portions of this region; for certain compositions and temperatures, mixtures with properties resembling those of the liquid alloys, or mixtures with properties characteristic of the alloys in the solid state predominate.

This recent work emphasizes still more the relationship between the form of the constitutional diagrams and the systematic variation in the casting properties of alloys with variation in their composition. They bring into sharper focus the more important aspects of this relationship, reflect the variations in quality occurring with gradual variation in composition, and indicate ways of improving the properties and the choice of the best alloys for the production of castings with predetermined properties. As a concrete example of the use of "composition — property" diagrams in solving industrial problems, the author would refer to his choice of alloys of the Al–Si system, instead of alloys of the Al–Cu system, which have not proved their worth, for the production of castings required to possess a certain hydraulic resistance [85]. The successful solution of this problem was predicted by the results of previous work carried out by A. A. Bochvar and Z. A. Sviderskii, who discovered the relationships governing the variation in hydraulic resistance of the above-mentioned alloys [86].

It must be mentioned that the solution of the general problem of the development of a theory of the casting properties of alloys is impeded by the difficulties of studying such properties of liquid metals and alloys as viscosity, surface tension, absorption and evolution of gases and so forth. It is merely necessary to point out, for example, that the relationship governing the variation of surface tension of alloys has been studied for 15 binary metal systems in all, and then mainly for systems of very fusible metals [79]. The variation of the viscosity of alloys with variation in composition and the form of the constitutional diagrams has been studied on approximately the same scale [41, 88 - 91]. This information on the properties of liquid alloys can be used in developing the theory of the casting properties of alloys only for explaining isolated facts, for example in studying the fluidity of alloys.

In the complex of the casting properties studied by us, therefore, we included the construction of experimental "composition — surface tension" curves of alloys and a search for surface-active additions for a number of alloys.

The combined study of the physicochemical properties of liquid metals and their casting properties is a matter for the future; it will enable us to discover the "fine" features of the solidification process of alloys of different compositions, differing in viscosity, surface tension, etc., and will give both a qualitative and quantitative expression to these features. The "fine" features of the solidification process of metals and alloys evidently also includes the nucleation of dislocations [238], which have a substantial bearing on the structure and properties of cast crystals, and of semiconductors in particular [239].

The foregoing review of the development of the theory of the casting properties of alloys shows that physical theories and physicochemical methods of research are being applied on an increasing scale to the solution of this problem. Despite their apparent remoteness from practical foundry work, these theories sometimes play a decisive part in the development of industrial methods of producing castings, for example, in the production of castings by the continuous casting method, in the production of precision castings by the investment process and so forth. An essential part in this has been played by the fact that the examination and improvement of the progressive technique of continuous casting involved a study of the equilibrium (and nonequilibrium) of metal alloys in the liquid-solid state and the theory of solidification.

The successful solution of such problems is assisted by the construction of "composition — property" diagrams, giving the quantitative variation of the property essential for use under the influence of composition and other external factors. It is obvious that the development of ideas on the structure and properties of metals in the liquid state,

on the results of alloying and on the casting properties of alloys is impossible without a study of the "composition —
surface tension," "composition — shrinkage," "composition — hot-shortness" and similar diagrams.

Although these diagrams cannot provide an answer to every problem of foundry practice, their experimental
construction nevertheless forms a necessary stage in the provision of scientific bases for foundry practice.

The main attention of this monograph is devoted to experimental evidence of the regular variation in the casting properties of alloys as a function of their composition and the form of the constitutional diagrams of binary metal systems. In the first place, obviously, an attempt is made to apply these laws to alloys of ternary systems and also to find a connection between such properties of alloys as surface tension and fluidity, surface tension and modi-
fiability, crystallization of alloys and fluidity and so forth.

The writer endeavored to obtain the data for the construction of "composition — property" diagrams under conditions which permitted the influence of external factors to be eliminated, and to bring out more fully the physical
meaning of the phenomena, without resorting to already existing concepts and hypotheses. The principal objective
of the work was to discover new aspects of the general theory of alloys, to provide theoretical reasons for the various
stages in foundry technique, and to indicate the possibility of predicting and sometimes calculating with sufficient
accuracy the value of the property sought, and of knowing the direction and magnitude of its variation with variation
in the composition of the alloys and certain external factors.

The author believes that the principal character of the phenomena and laws he has observed in the case of
aluminum or copper alloys may be generalized and applied to the solidification of steel, cast iron, and other alloys.

This is shown by the agreement between the main conclusions reached in a study of shrinkage and hot-shortness
as characteristics of foundry alloys, and observations on the cracking of welds when filler metals of various compositions are used. These results are given in the corresponding chapters of the book.

Chapter III

SURFACE TENSION OF LIQUID METALS AND ALLOYS

Surface phenomena play an important part in the metallurgical processes of the production of steel and other alloys and in the development of alloy theory. The surface tension at the boundary of the liquid metal with gases or solid crystals and its variation during melting, oxidation, and crystallization has an important influence on the kinetics of the processes and on the quality of the melted metal. It expresses the surface activity of the components of solutions, and their ability to concentrate in the surface layers at the liquid metal — gas, metal — slag, metal — crystal interfaces. At low concentrations of the dissolved substance, this activity is expressed by the difference in the value of the surface tension of the pure solvent and the solution at the boundary of separation of the phases. Within the limits of this boundary there may be formed a layer consisting solely of atoms or molecules of the surface-active component of the solution.

The practical significance of research on the surface tension of liquid alloys is also evident in the examination of modifying processes for example of iron (with manganese), silumin (with sodium), steel (with boron) and so forth, resulting in a sharp improvement in structure, and the mechanical, casting and other properties of the alloys.

In recent years, in connection with the development of special forms of casting metal in a variety of molds, and particularly in the production of precision castings of fine cross sections and complicated contours, the question of the influence of surface tension of the liquid metal on the properties of the castings has attracted the attention of investigators to an increasing extent.

Interest in the surface phenomena of liquid metals is also increasing in connection with the development of forms of machines, such as mercury boilers, in the tubes of which mercury heated to a temperature of 400 - 600°C is continuously circulated, or steam turbine superheaters, through which are pumped metal heat carriers — alloys of lead and bismuth, potassium and sodium, and others, heated to 700 - 800°C. The surface properties of these alloys play an important part in devices for maintaining a constant temperature in various units of equipment, for example reactors, certain parts of melting furnaces, molds for metal or glass castings, in the transfer of metals, for example, aluminum alloys through pipes from the melting furnaces to the refining vessels, mixers and molds of continuous casting machines, during flow in channels and molds in the production of castings and the like. The proper carrying out of the above enumerated processes for the production and utilization of liquid metals, the improvement and intensification of these processes are bound up with the need for extensive research on the properties of metals and alloys in the liquid state, and in particular research on surface phenomena.

The significance of these investigations for the development of the theory of metallurgical processes and alloy theory is that they make it possible to disclose in greater detail the structure of liquid melts and reveal the features of the behavior of the components in solution and their influence on the variation of the free energy of the solutions.

In their turn, these variations have an influence on the growth of the crystalline phase during solidification and on its shape and size, which to a considerable degree is reflected in the mechanical and other (service) properties of metals and alloys. The presence of surface-active impurities in melts results in their adsorption on the boundaries of the crystallizing phase and in the reduction in size of the latter; at the same time, useful deviations from equilibrium are also possible, for example, an increased tendency of the solutions to supercooling. The adsorption layers of surface-active substances formed during crystallization have a considerable influence on the results of subsequent treatment processes, in particular, in the annealing and tempering of steels and alloys, as well as in recrystallization. Apart from this, the increase in size of the crystals in the sintering of powders or when a polycrystalline metal, which is free from surface-active additions, is heated, and the rounding of the corners and edges of crystals when heated to temperatures close to the melting point, are due to surface forces resulting from lack of symmetry in the atomic structure of the surface of crystals.

Research in the field of surface phenomena and surface tension of liquid metals is being conducted in the following directions:

A. Theoretical work on the determination of the value of surface tension on the basis of molecular-kinetic and other concepts of the structure of metals and alloys in the solid and liquid states, and the reactions of atoms and electrons in the body of the metal and on its surface. In this group, a considerable amount of work has been done to show the connection between surface tension and physicochemical properties, such as atomic volume, boiling and melting points, heat of fusion, emission and so forth.

B. Theoretical and experimental work to determine the dependence of the surface tension of metals and alloys on composition and temperature, and also to determine this dependence in relation to the form of the constitutional diagram. It is also possible to include here work showing the connection of certain physical and mechanical properties and the structure of alloys in the solid state with the variation in surface tension of the liquid alloys.

C. Experimental work on the study of surface phenomena, adsorption and desorption in the metallurgical processes of oxidation, reduction, and modification, as well as in welding, tinning and the like.

The following is a brief account of some of such work having a direct bearing on the problem studied.

Theoretical Work on Calculating the Surface Tension of Metals

Liquid substances have the well-known characteristic property of contracting their surface, as a result of which small droplets of water, mercury, and molten metals assume a spherical shape. This is because forces tending to prevent the drop from spreading act on the surface of the drop. These forces compel the drop to rise to a higher level in narrow tubes than in wide ones. The origin of these forces is related to the fact that the particles of substance situated on the surface of a liquid are acted on by the forces of attraction of their neighbors situated below them and at their sides; from the direction of the free surface of the liquid, these forces are absent, while inside the liquid they are in mutual equilibrium.

Such an arrangement of the particles and their interaction gives rise to the formation of free surface energy, one form of which is surface tension. At the same time, each particle of substance situated on the surface of the liquid experiences the attraction of the underlying particles, which is directed to the interior of the liquid and is perpendicular to the surface.

The phenomenon of the mutual solubility of two different liquids is also related to the character and magnitude of their surface energy: the greater the mutual solubility of the liquids, the lower the surface tension at their interface. An increase in temperature, by producing a reduction in surface tension, increases the mutual solubility of the liquids.

The property of the surface of a liquid due to the interaction of the molecules and atoms in the interior and on the surface of the liquid cannot be compared with the property of a surface which is due to the action of a resilient film situated on the surface of a liquid, for example a film of oxides.

The theory of the surface tension of liquids and of liquid metals in particular is directly related to the problem of the liquid state of a substance and is developed on its basis.

The study of the surface tension of liquids of different characters has been developed most and been most successful in the field of organic substances, in which the orientation of complex molecules on the surface is a normal condition. In this branch of chemistry, extremely interesting results have been obtained in regard to the methods and accuracy of surface tension measurements, and also in regard to the use of such results in revealing the physical structure of substances and solutions differing in composition, in determining the behavior of the molecules in the surface layer, in ascertaining the temperature and concentration relationships, and so forth.

Academician V. A. Arbuzov in one of his recent articles on this subject [92] writes of the surprising agreement between the calculated and experimental values of surface tension and the parachor for individual groups of organic compounds, permitting the determination of standard values of the constants for members of homologous series of hydrocarbons and their surface tension. Thus, the surface tension of benzene, calculated from the parachors of carbon, hydrogen and their bonds, is found to be 207.1 d[ynes]/cm, while experiment gives 206.2 d/cm [93]. The values of the surface tension and parachors of organic compounds are thus a reliable criterion of the classification of the compounds and the discrimination of their structural formulas.

Methods of determining the surface tension of liquids which wet the walls of capillaries are distinguished by their diversity and high accuracy [94]. They include methods for determining this property in microscopically small volumes of liquid [95].

Matters are different when we come to study the properties of liquid metals and alloys. The theory of the problem has been inadequately developed, the methods of determination are inexact and the experimental results scanty. As pointed out by V. K. Semenchenko, this is largely due to the difficulty in working with liquid metal solutions [33]; this difficulty has not yet been surmounted.

The following is a brief review of the results of work on the theory of the surface tension of metals.

A theory of the surface tension of metals was first advanced by Ya. I. Frenkel' in 1918 [28]; according to this theory, surface tension is regarded as the electrostatic energy of a double electric layer formed on the surface of the metal by the fact that the electronic gas leaving the metal has a different energy from the energy of the electrons in the metal.

This theory was subsequently developed by Ya. G. Dorfman [96], who calculated the energy of the double layer on the basis of modern quantum mechanics and showed that the calculated and experimental values for the surface tension of metals are of the same order.

According to the views of V. K. Semenchenko, the very high values of the surface tension of liquid metals, and in particular the surface activity of alkali metals dissolved in mercury, compared with other surface-active substances in solution, is due to the presence in metals of free electrons, which are absent in other substances. In the opinion of V. K. Semenchenko, the electron theory of metals does not reflect in its formulas the function of ions or nuclei, on which the individual properties of metals depend; he has therefore applied to metals the concept of the so-called generalized moment, an energy characteristic expressing the energy properties of the ion

$$m = eL : r,$$

where e is the charge of the electron, L the valence of the ion, and r its radius. He furthermore showed that the activity of additions dissolved in alloys (solutions) increases in parallel with the difference between the generalized moments of the atoms of solvent and dissolved metal. If the generalized moment of an atom (molecule) of the solvent is greater than the generalized moment of an atom of the dissolved substance, the activity of the addition is negative, and the value of the surface tension of the solution increases; if, on the contrary, the moment of an atom of the addition is large in comparison with the moment of an atom of the solvent, the surface tension of the solution decreases. This position of the theory, based on the generalized moments of metals, is shown in Fig. 16 by curves of surface tension of amalgams versus their composition. With increase in the generalized moments of the alkali metals (increase in the radius of the ion) from lithium to cesium, their influence on the surface tension of mercury becomes greater; an addition of cesium to mercury in an amount of 0.1 at.% practically halves the surface tension of the mercury [33].

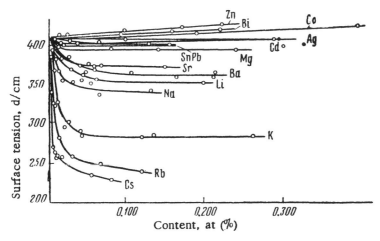

Fig. 16. Dependence of surface tension of amalgams on composition.

Subsequent work, carried out with a view to developing the molecular theory of surface phenomena in metallic solutions, and proposed by V. K. Semenchenko, showed the positive aspects of this theory, introduced corrections in it and endowed it with fresh experimental data on the surface tension of metals.

In 1945 - 1946, two articles by A. G. Samoilovich were published on the theory of the surface tension of metals [100], based on a consideration of the properties (density) of the capillary surface layer of metals. The author proposed an equation for calculating the surface tension of metals, which gives values agreeing with the experimental values. In 1947, A. G. Samoilovich [101] defined more exactly the physical essence of the problem of the nature of the forces giving rise to the surface tension of metals. In his opinion, the electron gas distributed in the interior of the metal with uniform density gradually loses the latter as the surface is approached. At the air-metal interface, the electron gas on penetrating the potential barrier produces "something of the nature of an electric double layer" on the surface of the metal. The resulting reaction between the forces of this layer and the surface of the metal tend to cause the free surface of the metal to contract; from the condition of the equilibrium of the forces, the author derives a formula for calculating the value of the surface tension of metals.

TABLE 1. Value of Surface Tension of Metals Near the Melting Point, Obtained Experimentally and by Calculation According to the Data of Various Authors, Text Books [79, 259], and Reference Books [135, 163]

Metal	Value of surface tension, d/cm		
	Experimental	Determined by the author (experimentally)	Calculated
Ag	916 [115], 923 [135, 259]	—	748 [100, 101], 810 [102], 940 [103], 903 [79] *
Al	494 [130], 520 [131], 502 [132], 349—412 [133], 900 [118], 840 [136]	860±20	829 [102], 637 [103], 668 [79]
Au	1125—1128 [115, 135, 259]	—	814 [102], 1042 [103], 926 [79] *
Bi	375—390 [259], 343 at 770° [115], 368 [99], 390 [79]	380±10	363 [96], 396 [103]
Cd	564—637 [259], 564 [135]	550±10	1156 [104, 101], 880 [102], 517 [103]
Cu	1115 [115], 1178 [131], 1060 [79]	1180±40	500—600 [96], 1128 [100, 101], 1100 [103], 1157 [79]*
Ga	358 [79], 735 [135]	725±10	—
Hg	460 [33], 402—515 [259], 473 [115], 465—470 [79]	500±15	500—550 [96], 644 [100, 101], 530 [103], 797 [79]
K	400 [130]	—	224 [100, 101], 146 [103], 91[79]
In	340 [135]	—	—
Mg	563 [130], 542 [118]	530±10	433 [103]
Na	191—253 [259], 205 [130]	—	50—60 [96], 400 [100,101], 223 [103], 206 [79]
Ni	1320—1730 depending on medium and support 1570 [259]	—	1334 [79]
Fe	1050—1200 [120], 1450 [259], 1210 [79], 880—1150 [121]	—	1180 [103], 1210 [79]
Pb	441—470 [259], 442 [135], 423 at 750° [115], 450 [79]	410±5	463 [79], 448 [103]
Sb	349—383 [259], 383 [135], 368 at 750° [115]	395±20	—
Se	105 [79]	—	—
Sn	523—652 [79, 259], 526 [135], 624, [118], 530 [79]	525±10	660 [79], 698 [103]
Te	436—485 [79], 401 [135]	—	—
Zn	550 [115], 753, 816 [259], 800 [118]	750±20	300—400 [96], 1397 [100, 101], 748 [103], 726 [79]

*Calculations by other authors give values 2 or 3 times higher than these [79, 259].

In 1949, an article by A. E. Glauberman was published on the theory of the surface tension of metals [102], in which a critical review is given of the work of A. G. Samoilovich. The author shows that the decisive part in the surface tension of metals belongs to the excess potential energy of the surface particles, and not to the kinetic energy of the electrons, as considered by A. G. Samoilovich. The mechanism of the surface tension effect is that due to lack of neighbors in the boundary layer, surface energy, which is a result of the crystal structure of metals, is produced in the surface particles. This energy is related to the electrostatic (coulomb) forces of mutual attraction of the ions of the metals.

A. E. Glauberman does not exclude the influence of the kinetic energy of the electrons on the surface tension of metals, but regards them as forces of repulsion, ensuring equilibrium of the crystal lattice. They constitute half the value of the surface tension calculated from the potential energy produced by the forces of atomic attraction.

This treatment of forces giving rise to surface tension is based on the theory of the metallic state and the quasicrystalline structure of liquids.

A. E. Glauberman gives the following equation for calculating the surface tension of metals of the cubic system with a face-centered lattice (Cu, Ag, Al and others):

$$\sigma = \frac{(2e)^2 \, 0{,}0074}{2a^3 \left[1 - \frac{2\pi}{3} \left(\frac{R_i}{a} \right)^3 \right]^2},$$

and for metals with body-centered lattice (K, Na, Fe and others):

$$\sigma = \frac{(2e)^2 \, 0{,}0087}{a^3 \left[1 - \frac{2\pi}{3} \left(\frac{R_i}{a} \right)^3 \right]^2}.$$

In these formulas, e is the charge of an electron, R_i the radius of an ion, and α the interatomic distance of the metals. The surface-tension calculations made by E. A. Glauberman show agreement between the calculated and experimental values.

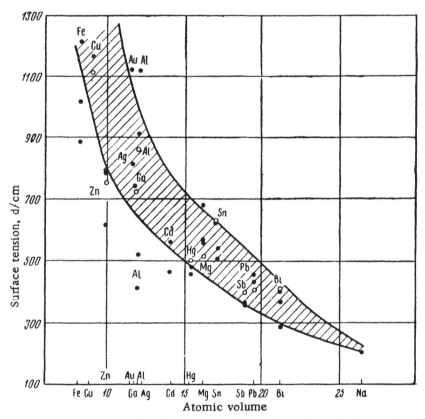

Fig. 17. Dependence of surface tension of liquid metals on their atomic volume: ● Data of various authors; ○ experimental data of our laboratory.

To conclude this review of previous work, Table 1 shows comparative values for the surface tension of metals obtained experimentally and by calculation using the formulas and equations proposed by various authors. The table also includes values of the surface tension of metals obtained on the basis of empirical formulas relating surface tension to the physical properties of metals. Included among them are data obtained from Stefan's formula [96], indicating the connection between surface tension and heat of evaporation, from S. N. Zadumkin's formula [103], connecting the surface tension and the atomic volume of metals, and L. L. Kunin's formula, who calculated the surface tension of metals, using the value for the work of exit of an electron [79, 104]. L. L. Kunin showed that the surface tension of a number of metals is a periodic function of the atomic number.

The data given in Table 1 indicate that the best agreement between the experimental and calculated data for a number of metals is found by using S. N. Zadumkin's formula ($\sigma = \alpha \cdot D/A$, where D is the density, A the atomic weight, and α is a coefficient equal to $7.87 \cdot 10^3$ erg/cm^2 and L. L. Kunin's formula ($\sigma = 444.5\psi/R^2 = 110$ erg/cm^2, where ψ is the work of exit of the electrons and R the radius of the atom), which have been proposed in recent years. The formulas of these investigators differ from those of other authors in their simplicity.

Figure 17 gives a diagram showing the connection between the surface tension of metals and their atomic volume. We have constructed the diagram on the basis of recently published data and our determinations of the surface tension of a number of metals. The values for the atomic volumes of the metals take into account the variation in density on heating and on transition to the liquid state. It can be concluded from these data that the greater the atomic volume of a given metal, the lower is its surface tension.

In addition to the factors mentioned above, we also consider it possible to use an energy factor, such as the standard value of the entropy of a metal. A comparison of these two properties shows that the higher the entropy, the lower is the surface tension. This is in agreement with the fact that according to O. Kubaschewski [258], the standard value of entropy characterizes the decrease in orderliness of the atoms.

Theoretical and Experimental Investigations of the Surface Tension of Alloys as a Function of Composition and the Form of Constitutional Diagrams

Investigations of "composition — surface tension" ("composition — σ") diagrams currently cover only 15 - 20 binary metal systems, and these are far from being complete. These investigations on the whole relate to alloys of fusible systems and amalgams, for which certain relationships have been observed in the variation of their properties. The data obtained for alloys are of course of limited significance and the "composition — σ" diagrams cannot yet be regarded as definite. The theories proposed for the surface tension of alloys and the views expressed on the character of the chemical reaction between the components in them, and the structure of liquid alloys in relation to the variation in surface tension are also not entirely definite, since there is conflict between the theoretical views and the actual "composition — σ" diagrams.

The general relationships of the variation in surface tension of aqueous and other solutions as a function of the composition and the form of the binary phase diagrams have been examined in papers by N. A. Trifonov and his collaborators [105 - 108]. In the first of them [105] formulas are given for estimating the surface tension of two-component nonmetallic systems, proposed by various authors, commencing with 1882, and three fundamental types of "composition — σ" curves are examined for binary systems, the components of which form a chemical compound. A maximum, minimum or transition point on the surface tension curves may correspond to this compound. A maximum is evidence of a strong chemical reaction between the components, and a minimum corresponds to the decomposition of the molecules of the system [106, 107].

Analysis of the "composition — σ" curves shows that they have no maximum if there is a considerable difference in the values of the surface energies of the components. The form of these curves also depends on the degree of dissociation of the compounds, due to the temperature [105].

Certain relationships of the variation in the surface tension of melts of inorganic salts, in particular chlorides of the alkali metals, have been studied in work connected with the metallurgy of magnesium [109]. For the eutectic systems $NaCl-MgCl_2$, $KCl-MgCl_2$ and others, it has been shown that the "composition — σ" curves do not vary in accordance with the rule of additivity, the deviation from which is particularly marked in curves of the temperature coefficient of this property (Fig. 18). Figure 19 shows the isotherms of the surface tension of melts of the ternary system $NaCl-MgCl_2-KCl$.

A theory of surface tension for alloys was first proposed by V. K. Semenchenko as a theory of generalized momenta. In recent years, some limitations have been introduced into this theory. Thus, in work on the investigation of the surface tension of bismuth alloys [99], it was shown possible to use the theory of generalized momenta for determining the "effective" momentum of metals, which is a more complicated problem than calculating the "ordinary" momentum from the charge of the electron and the radius of the ion. Recent work on the investigation of thallium amalgams also shows the limitations in the application of the theory of generalized momenta to the assessment of "composition — surface tension" diagrams, since a minimum was found on these curves for alloys of the Hg—Tl system [110]. In connection with this, the authors assume that the theory of generalized momenta is applicable only to dilute solutions (melts). As will be shown below, the theory of generalized momenta in the form pro-

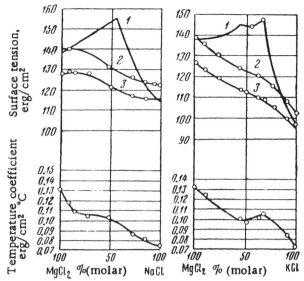

Fig. 18. Surface tension isotherms of binary systems having a chemical compound: 1) At the melting point; 2) at 700°C; 3) at 800°C.

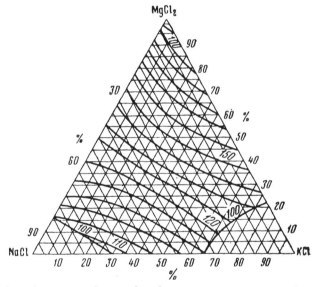

Fig. 19. Dependence of surface tension on composition of melts of chlorides in the system NaCl–MgCl₂–KCl.

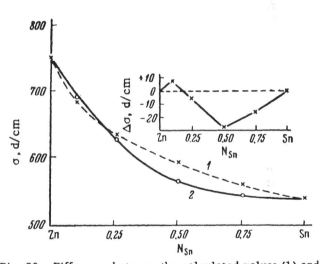

Fig. 20. Difference between the calculated values (1) and experimental values (2) of surface tension for alloys of the Zn–Sn system.

posed by V. K. Semenchenko fails to give directly satisfactory results when used for the assessment of surface tension of the alloys studied by us and based on aluminum [111] and zinc.

The possibility of constructing theoretical "composition −σ" diagrams for binary metal systems was shown in work by B. Ya. Pines on the basis of the energy characteristics of the atoms of alloyable metals or the heat of evaporation of the latter [112]. In this work, however, no comparison is given between experimental and calculated data. Nor are such data present in a paper by A. E. Glauberman and A. M. Muzyrchuk [113], who proposed formulas for calculating the surface tension of alloys of two metals with cubic crystal lattices. These formulas are based on concepts of the "effective charges" of the alloy metals.

An assessment of the surface tension of some binary systems of liquid metal alloys and a comparison of data obtained analytically and experimentally was made quite recently by Taylor [114]. His method of calculating the surface tension of binary alloys is based on concepts of the activity of the components in the surface layer, the total free energy of the interfaces and the quasicrystalline structure of liquid metal solutions. Comparison of the theo-

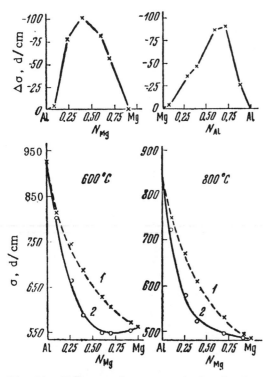

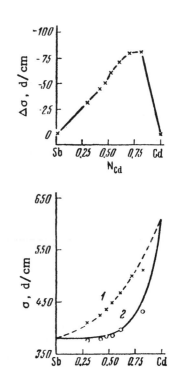

Fig. 21. Difference between calculated values (1) and experimental values (2) of surface tension for alloys of the Al–Mg system.

Fig. 22. Difference between calculated values (1) and experimental values (2) of surface tension for alloys of the Sb–Cd system.

retical and experimental data showed good agreement when six simple eutectic systems were considered; the maximum deviation did not exceed 30 d/cm for the Zn–Sn system, as is shown by Fig. 20. The best agreement between the data was found in the case of alloys of the Al–Sb, Pb–Bi and Sn–Bi systems. For alloys of the Al–Zn system, the calculated values were higher than the experimental values by 60 - 70 d/cm (for the 800 - 650°C isotherms and for alloys of high zinc content). Still closer agreement was found for Al–Mg, Zn–Sb and Cd–Sb alloys; these three systems are characterized by the presence in them of chemical compounds in the solid state. The maximum discrepancies between calculated and experimental values as a rule correspond to the concentrations of the alloys at which these compounds are formed (Fig. 21). This points definitely to a chemical reaction between the layers of atoms in the liquid state on the liquid-gas interface. For simple eutectic systems, for which there is no such reaction the analytical method gives good results, enabling one to predict the character of the variation of the surface tension of alloys with their composition.

The development of the theory of surface tension for alloys of binary systems in recent years has thus made it possible to indicate the relationships governing the variations in "composition − σ diagrams" and calculate the value of σ for the simplest cases of interaction of the components, for example, for alloys of eutectic type.

It should be pointed out that previous work in this direction had indicated the existence of some connection between the variation in surface tension and variation in composition. Thus, the work of F. Sauerwald, who was among the first to study the properties of liquid alloys, showed that the "composition −σ" isotherms of the Cu–Sb system exhibit a point of inflection corresponding to the composition of the chemical compound Cu_2Sb in the solid state [115]. No such points of inflection on the "composition − σ" curves are found for other systems of alloys, for example Cu–Ag, in which no chemical compounds are formed [116].

Similar results were obtained by Greenaway in a study of alloys of the systems Cd–Sb and Pb–Sb [117. The first of these is characterized by a chemical compound and the curve of variation of surface tension of the alloys with composition shows an inflection, corresponding to the composition of the compounds CdSb or Cd_3Sb_2 (Fig. 22). For Pb–Sb alloys, which form a simple eutectic system, the surface tension varies according to a smooth curve.

A direct relationship between the surface tension isotherm and the constitutional diagram was observed in the work of E. Pelzel; thus, in alloys of the Mg–Zn system, the chemical compound $MgZn_2$ is indicated by a minimum on the surface tension curve [118].

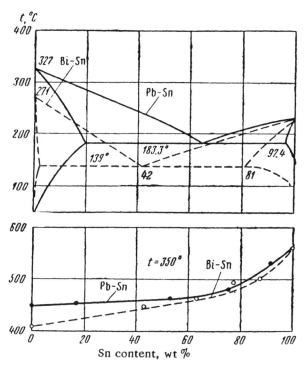

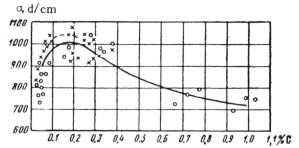

Fig. 23. Dependence of surface tension of Bi–Sn and Pb–Sn alloys on composition.

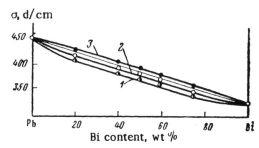

Fig. 24. Temperature and concentration relationship of surface tension of alloys of the Bi–Pb system 1) 300°C; 2) 400°C; 3) 500°C.

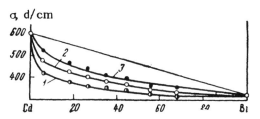

Fig. 25. Temperature and concentration relationship of surface tension of alloys of the Bi–Cd system 1) 300°C; 2) 400°C; 3) 500°C.

The diagrams of Figs. 23 - 25 show the results obtained by R. V. Bakradze and B. Ya. Pines in an investigation of the surface tension of four systems of alloys of eutectic type; the isotherms of this property have a monotonic character for these alloys, corresponding to cases of the absence of chemical reaction between the components of these systems [119].

The influence of carbon and oxygen on the surface tension of iron was determined in work done by B. V. Stark and S. I. Filippov [120]. They showed that carbon in a low content (up to 0.2 - 0.3%) raises and then lowers the surface tension of iron somewhat (Fig. 26); oxygen lowers it appreciably (Fig. 27). On the basis of these results, the authors calculated the magnitude of the combined adsorption of carbon and oxygen in steel and indicated the part played by it in the decarbonization processes of liquid steel.

T. P. Kolesnikova and A. M. Samarin [121] studied the variation in surface tension of Fe–Mn, Fe–Cr and Fe–V alloys as a function of the presence of ferrous oxide (oxygen) in them. Their work also indicated a connection between the surface tension of the alloys and the non-metallic inclusions contained in the latter.

T. F. Bas and G. G. Kellog [122] studied the dependence of surface tension on composition in the system Cu–S. They found that the addition of sulfur considerably reduces the surface tension of copper. Thus, with a content of 0.83% S, the surface tension is only 555 ± 7 compared with 1269 ± 20 d/cm for pure copper.

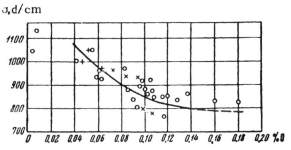

Fig. 26. Variation in surface tension of iron as a function of carbon content: O–Melt in atmospheric air; ×–melt under slag.

Fig. 27. Influence of oxygen on the surface tension of molten iron: O–For 0.01 - 0.03%C; × for 0.035 - 0.04%C; + for 0.06 - 0.08%C.

A more comprehensive review of the experimental data of the work mentioned above and other work on the study of the surface tension of alloys of binary systems is given in the book by L. L. Kunin [79]. It must be stated that there are no published data on the variation in surface tension of alloys of ternary systems.

Experimental Work on the Study of Surface Phenomena in Alloy Modification Processes

Investigations in this field attract attention in view of their considerable practical significance in the improvement of the structure and mechanical and other properties of alloys. It suffices to say that the results of work on the modification of cast iron by magnesium show that it is perfectly possible to double the strength of iron castings compared with the strength of castings made from iron which has not been modified; at the same time, the plasticity of the iron is sharply increased. A classic example of the improvement of the structure and properties of alloys is the modification of aluminum-silicon alloys by the addition of small quantities of sodium (in the form of sodium salts) to the liquid alloy, resulting in a sharp increase in fineness of the eutectic and in particular in an increase in the fineness of the silicon crystals in the eutectic and an improvement in the ductility of the alloy. It is only by this means that it has been possible to use Al–Si alloys on an extensive scale in machine construction, and in aircraft engine construction in particular.

The theory of the modification of alloys by surface-active additions and the mechanism of their action have been discussed by P. A. Rebinder [123], V. K. Semenchenko [33], Yu. A. Klyachko [124] and others.

P. A. Rebinder considers that an important part is played in the formation of crystals from the liquid metal by a stage of microheterogeneous condition, corresponding to the period of formation of crystals of maximum fineness (nuclei). The presence of surface-active substances in the liquid causes adsorptive films to be formed on the boundaries of the crystals, resulting in a reduction in the rate of growth of individual faces of the crystallization centers, and also in a variation in the conditions in which the crystals grow together or coalesce. Modifiers which may be used are surface-active metals (additions), possessing very small forces of cohesion in the liquid state, i.e., having a low value of surface tension, in particular metals of low melting point, and also some salts and oxides. The latter has been clearly confirmed by the work of S. M. Baranov, who made an extensive investigation of the part played by oxides of silicon ("high-silicon silicates") as surface-active component in steel, varying to a considerable degree the structure and properties of steel, both in crystallization from the liquid and in the transformations which take place in the solid state as the result of heat treatment [125]. P. A. Rebinder has also pointed out that the modifiers with the strongest surface-active effect must be substances which are not very soluble in the liquid from which they are adsorbed.

The author [111] found that low-solubility additions actually reduce the surface tension of the liquid metal (solvent), but not every addition of such a surface-active metal produces a modifying effect.

It is here appropriate to point out that, in accordance with investigations by M. V. Mal'tsev [126], lead and bismuth which, according to our investigations are surface-active in regard to aluminum and some of its alloys, exercise a modifying action and reduce the size of coarsely crystalline formations in some aluminum alloys.

Recent work by A. A. Bochvar and G. M. Kuznetsov on modification processes of eutectic alloys has shown that in the refinement of the structure of the eutectic of Al–Si alloys on the introduction of sodium, the decisive factor is the restriction in the growth of silicon crystals by a film of sodium silicide, which is soluble in the liquid alloy, but is insoluble in the solid silicon. The authors consider that the reduction in surface tension at the crystal-liquid interface and chemical adsorption do not explain the modifying action of sodium in silumin, and that, in this case, the essence of the process resides in a film inhibition of the growth of silicon crystals by a chemical compound of sodium and silicon which is insoluble in silicon and in the supercooling of the eutectic crystallization [127].

A systematic relationship has been found between variation (reduction) in surface tension and modification on the addition of calcium and boron to austenitic steel [79]. It was found that an addition of calcium in a quantity of 0.4 - 0.5% abruptly decreased the surface tension of the liquid steel (from 1800 to 1150 erg/cm^2), which had a favorable effect on the refinement of the steel structure. The addition of small quantities of boron (0.03 - 0.1%) to this class of steel also reduces the surface tension and is accompanied by a grain refinement of the steel and a reduction in the zone of columnar crystals.

We are not concerned here with other theories on the modification of alloys, for instance the theory which considers the part played by high-melting nuclei, discussed in detail in Chapter II. This theory does not relate the change in structure of castings, particularly the refinement of the primary structure, to the magnitude and variation of the surface tension of alloys. A detailed account of this theory will be found in the work of M. V. Mal'tsev [78, 126].

In the author's opinion, an essential point is the discovery of a mutual connection between the surface tension of a liquid metal and its casting properties, for instance fluidity. These results will be given in detail later. Here it will merely be pointed out that the connection between these properties was shown in work by O. S. Bobkova and A. M. Samarin [129] in experiments on chrome-nickel alloys. It was found that reduction in the surface tension of the alloys was accompanied by some increase in their fluidity. In a number of other investigations on the study of the properties of alloys, it has been shown that surface tension plays an important part in foundry practice, but as a rule no conclusive experimental data are given. In one investigation [128], it was shown that sulfur and phosphorus reduce the surface tension of cast iron and considerably increase the fluidity of the latter.

As previously pointed out, a study of adsorption phenomena plays a certain part in the development of the theory of solutions. A study of adsorption is of very great significance for metallurgists and metal researchers, since adsorption phenomena accompany the processes of melting, electrolysis, and refinement of metals. In the crystallization of metals, the formation of adsorption layers of impurities on the boundaries of the crystals and inside them, i.e., on the boundaries of the intradendritic cells and blocks is an important factor, and determines the structure and properties of the solidifying metal. The value of the surface tension is a characteristic feature and a criterion of adsorption from solution. This follows directly from the mechanism of adsorption and from the established concepts on the concentration of the dissolved substance at the interface between the liquid solution and a gas, another liquid or a solid.

The well-known Gibbs adsorption equation provides a possibility of calculating the excess of dissolved substance adsorbed on 1 cm² of surface from solutions of different concentrations:

$$\Gamma = - \frac{C}{RT} \cdot \frac{d\sigma}{dC}.$$

It follows from this equation that if $d\sigma/dC < 0$, i.e., if the surface tension decreases with increase in concentration, adsorption is positive ($\Gamma > 0$) and the dissolved substance is surface active, and vice versa.

Later (pp. 45-47), an assessment is given of the adsorptive capacity of some surface-active additions in relation to their effect on the variation in surface tension of aluminum, zinc and magnesium.

Methods of Determining the Surface Tension of Metals

There are various methods of determining the surface tension of metals, but the most common of them are the capillary rise method, the drop method and the maximum gas-bubble pressure method [33, 79, 94, 115].

In our investigations, we used the maximum bubble-pressure method which is employed most frequently of all for the determination of the numerical value of the surface tension of molten metals.

The principle on which the method is based is that of the introduction of gas into the molten metal, slowly, through a tube of small diameter (about 2 mm); the small gas bubble forming at the end of the tube (capillary) increases in size as the gas is supplied and the curvature of the bubble diminishes. When the radius of the gas bubble exceeds somewhat the radius of the tube, the bubble detaches itself from the end of the tube [93].

Equilibrium will correspond to equality of the diameters of the gas bubble and capillary. To increase the volume of the bubble by the small increment dV sufficient for its detachment, it is necessary to perform the work P·dV, where P is the excess of pressure above atmospheric. This work is expended in increasing the reserve of energy of the surface of the bubble in the liquid, equal to σ·dO, where σ is the surface energy and O is the surface. Thus,

$$P \cdot dV = \sigma \cdot dO.$$

For the case of a sphere

$$V = \frac{4}{3}\pi r^3; \quad O = 4\pi r^2; \quad dV = 4\pi r^2 \cdot dr; \quad dO = 8\pi r \cdot dr.$$

Consequently,

$$P \cdot 4\pi r^2 \cdot dr = \sigma 8\pi r \cdot dr$$

or

$$P_1 = \frac{2\sigma}{r}.$$

The pressure necessary for detachment of the gas bubble will also depend on the hydrostatic pressure P_2 to which the bubble is exposed and which corresponds to the specific gravity of the tested liquid and depth of immersion h:

$$P_2 = d \cdot g \cdot h,$$

where g is the acceleration due to gravity.

Thus,

$$P_3 = P_1 + P_2 = \frac{2\sigma}{r} + d \cdot g \cdot h.$$

The observed pressures P_3 at the moment of detachment of the gas bubble are related to the specific gravity of the manometer liquid d_m and the height of this liquid h_m in the manometer tube, i.e.,

$$P_3 = d_m \cdot g \cdot h_m$$

Consequently,

$$d_m \cdot g \cdot h_m = \frac{2\sigma}{r} + d \cdot g \cdot h,$$

whence

$$\sigma = \frac{rg}{2}(d_m \cdot h_m - d \cdot h).$$

For convenience of calculation, we express the last factor as the reduced height

$$h_\sigma = \frac{h_m d_m}{d} - h;$$

the value of the surface tension is then

$$\sigma = \frac{gr}{2} \cdot d \cdot h_\sigma \ \text{erg/cm}^2.$$

By way of example, the following is the calculation of the surface tension of commercial mercury. The height of the liquid in the manometer tube (measured repeatedly) for a capillary diameter of 0.21 cm and depth of immersion of the capillary 0.5 cm was 12.8 cm. The specific gravity of the liquid in our instrument (a saturated solution of common salt in water) was 1.14 g/cc. The reduced height

$$h_\sigma = \frac{12.8 \times 1.14}{13.55} - 0.5 = 0.57,$$

whence the surface tension of mercury

$$\sigma = \frac{981}{2} \times 13.55 \times 0.57 \times 0.105 = 464.31 \ \text{erg/cm}^2$$

Most reference works give the value of 485 ± 20 erg/cm^2 for the surface tension of mercury.

The apparatus used for the determination of the surface tension of molten metals is shown diagrammatically in Fig. 28. This is one of the forms of apparatus previously employed for this purpose [130].

The gas (argon) passes from a cylinder through a reducing valve into a rubber tube 1 and through the bottle 2, containing sulfuric acid, and the glass cylinder 3, filled with silica gel, and through a rubber tube into the quartz tube 5 filled with titanium turnings. The tube is situated in a furnace 6 which heats the titanium to 700°-750°C so that the latter absorbs the oxygen and moisture contained as impurities in the argon. After passing through the bottle 4, containing colored manometer liquid, the purified gas enters the capillary 7 immersed in the molten metal. As the argon bubble in the molten metal expands, the level of the liquid in the manometer tube gradually rises. The instant of detachment of the argon bubble is well observed visually by the abrupt drop in pressure. The maximum height of the liquid level observed before the detachment of the gas bubble is noted. The height of the manometer liquid can be read to an accuracy of up to 0.05 cm; as a rule, this accuracy does not exceed 0.1 cm. For successful operation, the working surface of the capillary on which the formation of the gas bubble takes place must be carefully treated (polished), and steps should also be taken to ensure that the capillary is mounted in a strictly vertical position.

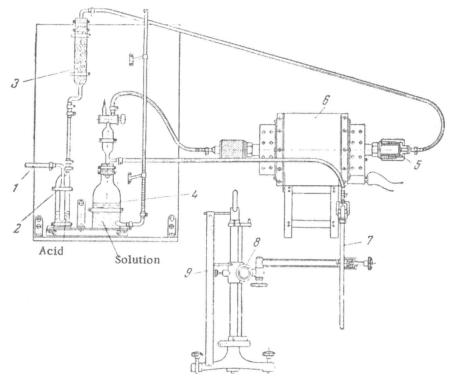

Fig. 28. Diagram of apparatus for the determination of the surface tension of metals and alloys.

For reading the depth of immersion of the capillary in the molten metal, this immersion being effected by means of a screw 8, an indicator is provided against a millimeter scale 9, permitting the depth of immersion to be read to an accuracy of 0.01 mm. As stated above, the value of the manometer pressure in centimeters is used in the calculations. The values of the specific gravity of liquid metals necessary for calculating the surface tension are obtained from reference sources, and for alloys they are calculated on the principle of additivity.

The values of surface tension for water, mercury, tin and others, obtained by means of this apparatus, agreed with well known values obtained from the literature and reference sources. It may therefore be considered that the method and apparatus adopted by us are perfectly suitable for the intended purpose. The simplicity of this apparatus may be found to be a definite advantage when it is used in works laboratories and under industrial conditions for the examination of liquid metal.

Determination of the Surface Tension of Aluminum and Some Other Metals

Published values of the surface tension of liquid aluminum at a temperature of 700°-750°C differ considerably from one another. In the earliest work by S. W. Smith [131] a value of 520 d/cm was found; then V. G. Zhivov [130], Yu. A. Klyachko [132], and S. V. Sergeev [133], using different methods, found it to be equal to 494, 502, and 349-412 d/cm, respectively. E. Pelzel in work done in 1949 [118] gives a figure of 914 d/cm; reference works, however, give 300 [163], 520 [135], and 840 d/cm; this value, found by A. Portevin and P. Bastien [136] relates to oxidized aluminum.

The value of the surface tension of zinc found by different authors using different methods is in better agreement than in the case of aluminum and varies within the limits of 700 - 800 dyne/cm. At 500°C, E. Pelzel [118] obtained $\sigma = 798$ d/cm while the temperature coefficient was 0.25 d/cm °C.

For the determination of the surface tension of aluminum, zinc and alloys of these metals, we used the method of the maximum bubble pressure described above; the molten metal was covered with a layer of flux of lithium and potassium chlorides. The quartz capillaries used had a diameter of 2 - 2.5 mm with a wall thickness of 0.3 - 0.5 mm. The aluminum and zinc had a purity of 99.99 and 99.97%, respectively.

The values of surface tension obtained for these metals are given in Table 2.

Repeated determinations of the surface tension of mercury using the apparatus described and capillaries having the diameters shown in Table 2 gave values of 500 ± 15 d/cm; the published data for mercury are 485 ± 20 d/cm. Thus, the accuracy of the determinations is quite acceptable for our purpose.

TABLE 2. Surface Tension of Aluminum and Zinc

No. of Experiment	Aluminum		Zinc	
	capillary diameter, cm	σ, d/cm	capillary diameter, cm	σ, d/cm
1	0.17	862	0.19	769
2	0.116	872.2	0.19	764
3	0.276	877	0.24	733
4	0.232	840	0.24	731
5	0.204	882	0.24	734.8
6	0.192	844	–	–
7	0.25	896.7	0.296	762
8	0.20	867.3	0.25	720
Mean		860 ± 20		750 ± 20

For copper, using articifial corundum capillaries, a value of 1180 ± 40 d/cm was obtained, which is in good agreement with previous determinations (1178 at 1150°C [131], 1120 at 1140°C [115], and 1060 at 1150°C [79]).

The values of the surface tension of tin, lead, and antimony given in Table 1 were also in good agreement with previous values. The purity of these metals was 99.9 - 99.99% (the last figure refers to antimony, which was subjected to double distillation).

These results made it possible to carry out extensive experiments on the determination of the variation in surface tension of alloys with their composition and the form of the constitutional diagrams.

Surface Tension of Binary Alloys, Based on Aluminum, Zinc, and Magnesium

A study of previous work on the determination of the dependence of the surface tension of binary alloys on their composition and the form of the constitutional diagrams shows that the majority of such investigations have been carried out on too few alloys in a given system, the concentration varying by 10 - 20% of the added component, as is shown by a review of this work [79].

When working on such a scale it is impossible to determine the most important part of the "composition – surface tension" curve corresponding to the small additions of the components to the solvent, that is to say, the part of the curve in which the essential feature of the phenomenon under examination may be largely revealed.

In contrast to the above-mentioned work, the author estimated the activity of the various additions to aluminum, zinc, and magnesium at concentrations commencing at thousandths and hundredths of a percent up to 10 - 20%. The temperature of the liquid alloys was 50 - 80°C above the liquidus temperature. The results of the observations are presented in the form of "composition – surface tension" diagrams; in some cases, these results are compared with the constitutional diagrams of the alloys considered.

The curves given in Fig. 29 show rather clearly that in relation to aluminum, surface-active metals are certain alkali and alkaline-earth metals (lithium, calcium, and magnesium), as well as elements of groups IV and V of the periodic system (tin, lead, antimony, and bismuth). The introduction of small additions of some of these components in the limits of thousandths and hundredths of a percent has quite an appreciable influence in reducing the surface tension. The slope of the initial part of the curve also shows more particularly whether the added component is surface-active in relation to the metal (solvent), i.e., whether it is capable of being concentrated in a layer on the metal-gas interface.

Figure 30 shows "composition–σ" curves for alloys of the system Al–Zn; the curves correspond to different experimental temperatures, i.e., to different degrees of heating of the alloys above the liquidus line. All three curves show that alloying of aluminum with zinc results in a gradual decrease in surface tension of the alloys to a content of 70 wt. % zinc, corresponding to 50 at. %. An appreciable increase in surface tension is found for alloys containing 85% of zinc or more. On the zinc side, the curves show that aluminum is an inactive addition in relation to zinc, since the surface tension of the latter is increased by the addition of aluminum.

The occurrence of a minimum on the "composition –σ" curve for this system may be due to the reaction of the components, which according to an earlier constitutional diagram is characterized by the occurrence of the chemical compound Al_2Zn_3 (78.4 % Zn) formed by peritectic reaction.

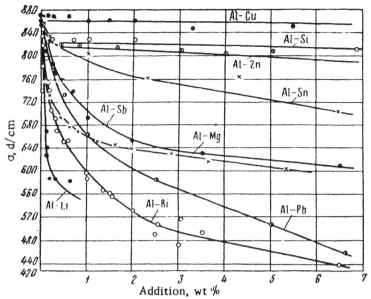

Fig. 29. Dependence of surface tension of binary alloys of aluminum on the content of the second component (addition).

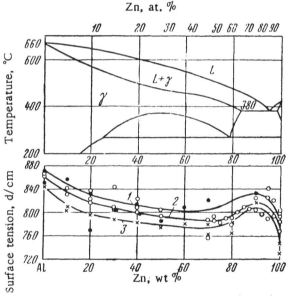

Fig. 30. Dependence of surface tension of Al–Zn
alloys on composition and temperature: 1) Heated
50°C above the liquidus; 2) heated 100°C above
the liquidus; 3) heated 150°C above the liquidus.

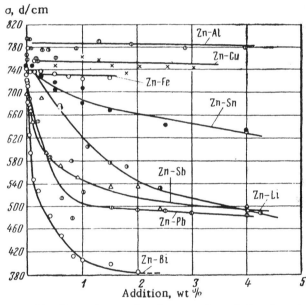

Fig. 31. Dependence of surface tension of zinc-based
binary alloys on the content of the second component
(addition).

E. Pelzel [118] also observed a similar increase in the surface tension of zinc by the action of aluminum, but
only at relatively low temperatures (500°C); at 800°C, this increase was no longer observed. J. Taylor, generalizing
these data theoretically, and comparing the Al–Zn system with the Sb–Zn and Cd–Sb systems, in which the reaction
of the components gives rise to the formation of definite stable chemical compounds and an abrupt change in trend
of the "composition − σ" curve, considers that the Zn–Al system is intermediate between the Zn–Sb and Cd–Sb systems with a pronounced chemical reaction between the components, and the Zn–Sn system, having a simple eutectic
transformation.

Systematic observations of the variation in surface tension of aluminum when alloyed with copper, silicon,
iron, nickel, chromium and silver failed to show any appreciable change in the value of σ for pure aluminum.
Evidently, these metals which are the most widely used in alloying additions in industrial aluminum alloys cannot
be regarded as surface-active in relation to aluminum.

In an earlier investigation [137], it was found that when aluminum is alloyed with copper and silicon, the surface tension is decreased, while when aluminum is alloyed with magnesium, the surface tension is increased. Numerous experiments by the writer have shown that the introduction into aluminum of copper or silicon in the limits 0.001 - 0.1 % and up to 10% showed no appreciable change in surface tension; the same additions of magnesium to aluminum in every experiment without exception resulted in a regular reduction in surface tension, as shown in the diagram of Fig. 29.

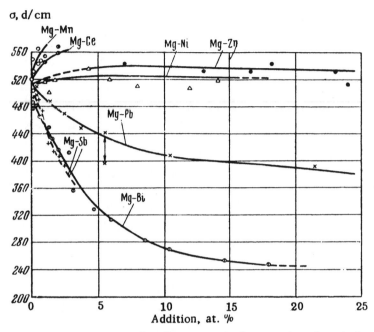

Fig. 32. Dependence of surface tension of magnesium-based alloys on the content of the second component (addition).

The diagram of Fig. 31 shows the results of an investigation of a family of zinc-based alloys. They are similar to the results obtained for aluminum-based alloys. In the present case, the most active components were the same metals as for aluminum; iron and copper are inactive in relation to zinc.

In alloys with magnesium, these same metals were found to be surface-active, the degree of reduction in the surface tension of magnesium when the latter was alloyed with bismuth and antimony being greater than when it was alloyed with lead, as shown by Fig. 32. When magnesium is alloyed with manganese, cerium, nickel, zinc and aluminum (principal additions in industrial magnesium alloys), the surface tension of the alloy is unchanged.

When the data obtained on the variation in magnitude of the surface tension of binary alloys are analyzed, it cannot be regarded as accidental that some alloying metals (lithium, antimony, lead, tin, and bismuth) have the same effect in reducing the surface tension of the three different metals (solvents) — aluminum, zinc, and magnesium. It appears that if the atomic volumes of the solvent and dissolved metal differ considerably, the dissolved metal is surface-active; if, however, the atomic volumes are near to each other, the added metal is inactive in relation to the solvent (Table 3).

The data given in Table 3 show that metals having a relatively large atomic volume are surface-active, while those having relatively low volumes are inactive in relation to aluminum, zinc, and magnesium.

Table 4 gives data characterizing quantitatively the relationship between the difference in the atomic volumes of aluminum and a number of alloying metals, as well as the reduction in the value of σ, referred to one atomic percent of the alloying metal.

It will be seen that the larger the atom of the metal introduced into the solvent, the greater is its effect in reducing the surface tension of the solvent. It is only necessary to introduce 0.1 at. % of lead or bismuth into aluminum for the surface tension of the latter to drop by 200 - 250 d/cm or by 20 - 30%. It is evident that the ability of these atoms to be adsorbed and to diffuse in the melts is very high, and that their concentration on the gas-metal interface proceeds rapidly and completely.

TABLE 3. Atomic Volumes of Liquid Metals and Their Surface Activity in Relation to Aluminum, Zinc, and Magnesium

Solvent-metals	Dissolved Metals	
	Surface-active	Inactive
Aluminum 11.40	Calcium 25.85 solid	Copper 7.94
Zinc 9.48	Lithium 13.9	Iron 8.2
Magnesium 15.7	Bismuth 20.91	Manganese 7.52 solid
	Lead 19.50	Nickel 7.67 solid
	Antimony 18.9	Silicon 11.7 solid
	Tin 17.1	

TABLE 4. Influence of the Addition of Metals of Different Atomic Volumes on the Variation in Surface Tension of Aluminum *

Added metal	Atomic volume	Difference in volumes	Reduction in σ for 1 at. % addition d/cm	Difference in the generalized moments $(M_{Al}-M_{add})$, e.s.u. $\times 10^{-2}$
Al	11.40	−	−	M_{Al} = 25.26
Bi	20.91	9.5	450	0.24
Pb	19.50	8.1	375	13.86
				2.26
Sb	18.9	7.5	240	-13.44
				2.06
Li	13.9	2.5	254	19.11
Ca	25.88	14.48	180	16.24
Mg	15.7	4.3	170	12.96
Sn	17.1	5.7	140	− 0.74
				12.26
Cu	7.94	-3.46		20.26
				15.26
Ag	10.27	-1.13	⎫	16.01
Zn	9.48	-1.92	⎬ Practically	13.69
Si	11.7	0.3	⎬ without	-22.97
Fe	8.2	-3.2	⎬ effect	−
Mn	7.52	-3.88	⎭	−

* Two figures in the last column of this table denote the difference in the generalized moments for different valencies of the given metal.

A similar nonuniformity in reaction of atoms of dissolved metals with atoms of the solvent may also be found in earlier work. Thus (according to the results of V. K. Semenchenko and N. L. Pokrovskii [33]), if the alkali metals are arranged in order of increasing atomic volume and in order of their action on the surface tension of mercury, the pattern shown in Table 5 is obtained.

The data of Table 4 show that the theory of generalized moments proposed by V. K. Semenchenko on the basis of the results of investigations of the surface tension of amalgams is not directly applicable to aluminum, zinc and magnesium alloys. Indeed, for an almost identical difference in the generalized moments of aluminum and of the metals of the first and second groups − copper, silver, zinc, magnesium, and calcium, only the last two appreciably reduce the value of aluminum, while the first two are practically without effect. Antimony and bismuth, which have a very small positive or even negative difference in moments, have a considerable effect on the reduction in σ. According to the theory of generalized moments, however, in the last case one ought to expect an increase in the surface tension of the solvent (aluminum), not a reduction, as actually occurs. The foregoing statements also apply to zinc-based alloys, since the atomic volumes of zinc and aluminum are very near together (9.48 and 11.4, respectively).

TABLE 5. Effect of Addition of Metals of Different Atomic Volumes on the Surface Tension of Mercury

Added metal	Atomic volume	Reduction in surface tension of mercury on addition of 0.1% alkali metal, d/cm
Hg	14.26	–
Na	23.7	95
K	45.5	150
Rb	56.2	190
Cs	71.0	210

It must be admitted that for the series of alloys investigated, the variation in magnitude of σ on alloying is more fully explained by the relationship of the atomic volumes of the alloyed metals than the relationship of the generalized moments. The high surface activity in alloys of additions with a large atomic volume is due to the fact that the value of the surface tension of pure metals itself depends on the atomic volume, as shown by the investigations of S. I. Zadumkin [103], and mentioned still earlier by S. W. Smith [131].

Of course, the volume factor cannot be regarded as the sole cause of the variation in surface tension of alloys on alloying. A leading factor affecting the variation in surface tension must be considered to be the separation of the components into layers in the liquid state.

If we consider the results obtained in regard to the connection between the variation in surface tension of binary alloys and the form of the constitutional diagram, we note that as noteworthy points on the "composition – surface tension" diagrams, we can indicate the points of practically complete suppression of the drop in the value of surface tension on alloying, such points corresponding to the monotectic points in constitutional diagrams, if the components of the alloys have limited solubility in the liquid state. This is shown by the diagrams in Fig. 33 for Zn–Bi and Zn–Pb alloys: approximately the same agreement is found between the "composition –σ" curves and the constitutional diagrams of the Al–Bi and Al–Pb systems, and according to other observations for Al–Cd alloys.

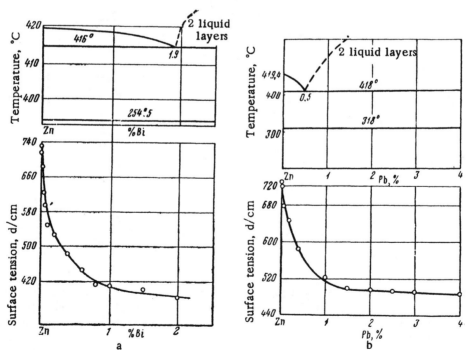

Fig. 33. "Composition –σ" curves compared with the constitutional diagrams for Zn–Bi alloys (a) and Zn–Pb alloys (b).

There is a definite interest in comparing "composition – surface tension" curves with the form of the constitutional diagrams of aluminum alloys, for example with metals of the fourth group – silicon, germanium, tin and lead (Fig. 34). These metals do not form chemical compounds with aluminum; the eutectic point in this system is shifted towards the component having a high atomic volume. In this series, the Al–Pb is the limit system, in which there is a spontaneous process of separation of the components in the liquid state. The "composition–σ" curves for Al–Si and Al–Ge alloys do not show any appreciable variation in σ with variation in composition, while for Al–Sn and Al–Pb alloys, a considerable variation in σ corresponds to the considerable difference in atomic volumes. In such cases, the constitutional diagram indicates a tendency of the liquid metal solvent to "expel" the metals dissolved in it, which naturally results in their separation and concentration at the metal–gas interface, and possibly at the crystal – liquid interface during solidification.

A second primary factor determining the variation in surface tension of liquid metal solutions must be assumed to be the presence in the system of a chemical reaction between the components, resulting in the formation of stable chemical compounds of the metals in the solid state. As shown by the diagrams in Fig. 29, a considerable reduction in the surface tension of aluminum is found to take place on the addition of lithium and antimony; the introduction of calcium and magnesium produces a lesser effect. This is evidently due both to the difference in the atomic volumes of aluminum and the added surface-active metals and to the circumstance that lithium and antimony form with aluminium high-melting chemical compounds with an open maximum, while alloys with calcium, at a certain concentration of the latter, form on solidification a chemical compound by peritectic reaction (i.e., melting with decomposition). The compound of aluminum with magnesium, although it has an open maximum, has a low melting point, which is below the melting point of the two components, and therefore complete dissociation of the compound in the liquid melt is possible.

Copper, iron, manganese, nickel, and chromium also form chemical compounds with aluminum but do not cause any variation in the value of σ on alloying. The inactivity of these metals is evidently determined by the size factor, i.e., their relatively small atomic volume.

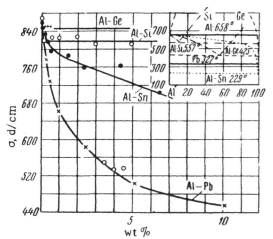

Fig. 34. "Composition – σ" curves and constitutional diagrams of Al–Ge, Al–Si, Al–Sn and Al–Pb alloys.

It should be mentioned that an examination of the surface tension of many series of aluminum-based and zinc-based alloys failed to reveal any connection between the variation of this property of liquid alloys and those elements of the constitutional diagram, which describe the character and magnitude of the mutual solubility of the components in the solid state. When components, such as zinc and silver, which are very soluble in solid aluminum, and copper, nickel, iron and others, which dissolve only slightly or to a very limited extent in solid aluminum at high temperatures, were introduced into liquid aluminum, no appreciable change in the surface tension of the liquid solutions was observed. On the other hand, antimony, which is practically insoluble in solid aluminum, and lithium, which is slightly soluble in it, sharply reduce the surface tension of liquid aluminum solutions. The statements to be found in the literature regarding the possibility of explaining the limited solubility of metals in the solid state on the basis of the values of their surface tension [79] are evidently only of limited significance or have no foundation in fact. This is all the more probable, since some of them are based on the surface-tension data of high-melting metals, for example, platinum, whose surface tension is very difficult to measure with sufficient reliability.

Thus, a systematic relationship between the variation in surface tension of alloys on variation in their composition and the form of the constitutional diagrams of the alloys is to be found quite definitely in systems where the components have limited solubility in the liquid state and to some extent in systems where chemical compounds of the two metals are formed, when the liquidus curves show a clearly pronounced maximum. Consequently, the surface tension of liquid alloys is quite definitely related to those regions of the constitutional diagrams which directly indicate transformations in the liquid solutions. This relationship is also observed in these cases if the region of liquid solutions is intersected by regions where there is a strongly pronounced chemical reaction between the components; this is in complete agreement with the views of N. S. Kurnakov on the occurrence of chemical reaction in solutions and the relationship of their properties with the form of the constitutional diagrams.

As was pointed out in the foregoing, earlier studies of the surface tension of alloys as a function of their composition and the form of the constitutional diagrams embraced a very small number of binary metal systems; ternary systems, however, had not been studied. The problem therefore arose of investigating the relationships governing the variation in surface tension of alloys in the aluminum corner of some ternary systems forming the basis of industrial alloys such as duralumin and silumin, as well as some zinc-based alloys.

Even our first observations showed that if a given alloying metal is surface-active in regard to the basic component of the binary alloy (aluminum, magnesium, or zinc), then it is also active in a ternary alloy on the basis of that component.

Figure 35 shows a typical "composition – surface tension" diagram for such a case. It indicates a regular drop in the surface tension of two binary alloys of the Al–Si system (with 5 and 10% Si) on adding lead to them. The action of lead in these alloys practically corresponds to its action on aluminum, as will be seen by comparing the curves of Fig. 35 with the similar curves for binary aluminum alloys given in Fig. 29. In both cases, the degree of reduction of the surface tension in the binary and ternary alloys with lead is practically the same. It follows from this that in the ternary solution Al–Si–Pb, there is no reaction between the alloying elements, silicon and lead, which would exert any appreciable influence on the form of the "composition – surface tension" curves of ternary alloys as compared with the curves for the binary alloys. The same quantity of lead introduced into the solution produces the same reduction in surface tension in both binary and ternary alloys; consequently, the adsorption of lead at the liquid–gas interface is the same, irrespective of whether the liquid is pure aluminum or a solution of inactive silicon and aluminum.

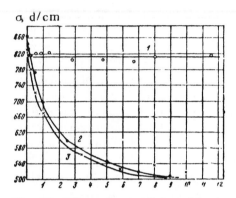

Fig. 35. Effect of an addition of lead on the variation in surface tension of alloys of aluminum with silicon: 1) Al–Si alloy (without lead); 2) alloy Al+10% Si; 3) alloy Al+5% Si.

A similar effect of surface-active lead and antimony is also observed in the case of the more complex, multicomponent alloy duralumin (Fig. 36). The surface tension of duralumin is lower then that of aluminum by 120 d/cm due to the presence of about 0.7% Mg in the alloy.

The results of a systematic investigation of the surface tension of ternary alloys of the systems Al–Si–Mg and Al–Zn–Mg situated on radial cross sections of the corresponding ternary diagrams are shown in Figs. 37 and 38. The form of the surface-tension curves of the ternary alloys with variation in their composition shows unmistakably that the reduction in surface tension of the alloys depends mainly on their content of the surface-active component, magnesium. Similar curves were also obtained for the system Al–Cu–Mg. Chemical reaction between copper, silicon, and zinc (inactive with regard

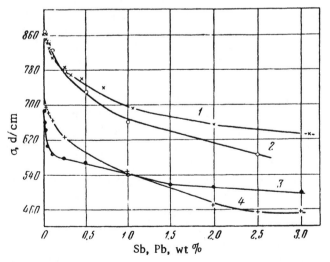

Fig. 36. Effect of antimony and lead on the surface tension of aluminum and duralumin: 1) Aluminum + Sb; 2) aluminum + Pb; 3) duralumin + Sb; 4) duralumin + Pb.

to aluminum) and magnesium in the liquid solution is either very slight or nonexistent, since it has no effect on the value of the surface tension of the solutions, which diminishes in proportion to the amount of magnesium contained in them. The formation of chemical compounds in these systems when the alloys pass into the solid state is not reflected to the "composition−σ" diagrams.

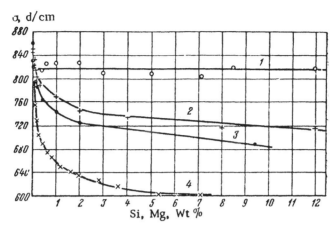

Fig. 37. Dependence of surface tension of alloys of the system Al−Si−Mg on their composition. 1) Al−Si alloys; 2) alloys Al−Mg: Si (1:1); 3) alloys Al−Mg$_2$Si (Mg: Si = 6.3 : 3.7); 4) Al−Mg alloys.

In the Al−Cu−Si system, variation in composition of the alloys and the transition to different phase regions of the diagram also did not result in any appreciable variation in the surface tension of the alloys. A similar conclusion is also confirmed by the results of experiments with alloys of duralumin type; the addition of the inactive components of this alloy − copper, iron, manganese and silicon (separately and together, adding them in turn) to liquid aluminum did not affect the surface tension of the alloys, while the addition of magnesium invariably reduced it.

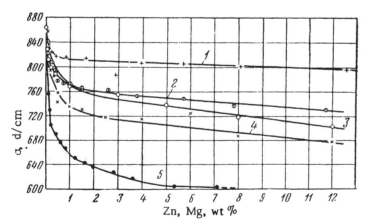

Fig. 38. Dependence of surface tension of alloys of the system Al−Zn−Mg on their composition: 1) Al−Zn alloys; 2) alloys Al−MgZn$_2$ (Mg: Zn = 1.6 : 8.4); 3) alloys Al−Mg: Zn (1:1); 4) alloys Al−Mg: Zn (3:1); 5) Al−Mg alloys.

Such variation in the surface tension of ternary and more complex alloys when alloyed with surface-active and inactive elements confirms the notion that the primary factor affecting the surface tension of alloyed metals is the relationship of the atomic volumes of solvent and alloying addition. If, in a ternary or more complex alloy, the addition has an atomic volume close to that of the solvent, the surface tension of the alloy is not appreciably altered, irrespective of the number of these additions. This relates both to cases where the addition does not enter into any definite chemical reaction with the solvent (for example, Al–Si alloys) and to cases where chemical compounds are formed (for example in Al–Cu, Al–Fe alloys, duralumin without the addition of magnesium, etc.). As soon as a metal with a large atomic volume, sometimes also not forming chemical compounds, is introduced into complex alloys, the surface tension of the alloy is sharply reduced. This explanation appears to be the most reliable for those cases where only one surface-active component is introduced into the solution. Its accuracy has also been confirmed by an investigation of the surface tension of a number of alloys of aluminum and zinc, into which several surface-active elements are introduced successively. If a less active element is introduced first, followed by a more active one, the surface tension of the solution is gradually reduced, as is shown by Table 6.

TABLE 6. Reduction in Surface Tension of Zinc by the Action of Several Surface-Active Components

Added metals (1 at. %)	Surface tension, d/cm
Zn	794.6
Zn + Sn	743.2
Zn + Sn + Sb	604.6
Zn + Sn + Sb + Pb	557.3
Zn + Sn + Sb + Pb + Bi	488.3

We may thus say that in this case, the effects of reducing the value of σ of the solution due to the addition of each metal are additive. If, however, the solvent is first alloyed with the component with maximum surface activity, i.e., having the largest atomic volume, successive additions of less surface-active metals (having smaller atomic volumes) have no effect on the value of σ of the solution; this is shown by the data of Table 7.

From the point of view of the adsorption theory, which is indissolubly connected with the surface tension theory, these results show that the occupation of the surface layer at the liquid–gas interface by metals of different surface activities takes place most rapidly and completely in the case of the addition of a metal having a relatively large atomic volume and limited solubility in the liquid-metal solvent.

Alloys in the aluminum corner of the ternary system Al–Mg–Pb were studied to ascertain the effect of chemical reaction of the components of a ternary solution on the variation in surface tension. In these alloys, the added surface-active components form the stable chemical compound Mg_2Pb.

As shown by the diagram in Fig. 39, the form of the curves in this case is more complex than for alloys of the ternary system Al–Cu–Mg and others. Here, the effect of reducing the value of σ of aluminum when the latter is alloyed is composed of the effect of both surface-active additions – lead and magnesium. The alloying of aluminum by two surface-active components forming a compound with each other (for a definite content of them) gives a greater effect (see curve 4) than the simple addition of the effects produced by each of them separately. A similar occurrence was also found in the case of several other alloys. The adsorption of the atoms of the alloying components on the liquid–gas interface in these alloys evidently occurs on their reaction, possibly in the form of molecules of Mg_2Pb, the volume of which may differ considerably from the volume of an atom of the solvent, which from our point of view is the principal cause of the variation in surface tension of solutions and melts.

TABLE 7. Surface Tension of Zinc and Aluminum under the Action of Several Surface-Active Components

Added metal (0.5 at.%)	Surface tension, d/cm
Zn	801
Zn + Bi	521.7
Zn + Bi + Pb	521.7
Zn + Bi + Pb + Sb	517.8
Zn + Bi + Pb + Sb + Sn	521.7
Al	868.7
Al + Bi	406.4
Al + Bi + Pb	414.7
Al + Bi + Pb + Sb	415.3
Al + Bi + Pb + Sb + Sn	410.3

The Adsorption of Surface-Active Metals on Aluminum and Zinc

The adsorption processes occurring in liquid-metal solutions receive close attention from investigators, since a study of these processes assists in revealing the mechanism of metallurgical reactions [120, 121, 138, 139] as well as of many other physicochemical processes [115, 123]. The existence of adsorption films at grain boundaries, and in view of some investigators also at the "block" boundaries, in solid metals has a considerable effect on the kinetics of structural transformations in heat-treatment processes [125], on the susceptibility of metals and alloys to intercrystalline corrosion [79] and so forth.

The direct study of the formation of adsorbed films on the surface of a liquid metal or at the solid phase-liquid interface involves considerable experimental difficulties. As a criterion which takes into account the adsorption films at the liquid metal – gas interface when a surface-active substance is introduced into the liquid metal, it is possible to use the value of the surface tension of the solution, since this is a characteristic manifestation of the forces of molecular or atomic interaction.

The experimental data quoted above concerning the regular variation of the "composition – σ" curves for alloys based on aluminum and zinc were utilized for determining the adsorption of some surface-active metals (additions) with the object of elucidating the structure of the surface film at the metal–gas interface. The dependence of the variation in the adsorption of the introduced component on its concentration in relation to the variation in surface tension of the solution is expressed by Gibbs's well-known equation:

$$\Gamma = -\frac{C}{RT}\left(\frac{\partial \sigma}{\partial C}\right)_{S,T}.$$

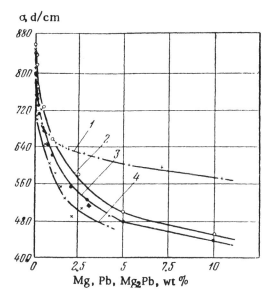

Fig. 39. Dependence of surface tension of alloys of the system Al–Mg–Pb on their composition: 1) Al–Mg alloys; 2) Al–Pb alloys; 3) alloys Al–Mg: Pb (1 : 1); 4) Al–Mg$_2$Pb alloys.

By making use of the method of graphical differentiation of the experimental surface-tension curve with variation in total concentration and determining the coefficient of surface activity ($\Delta\sigma/\Delta C$) for any concentrations, the variation in adsorption of the added component in the surface layer may be found as a function of its concentration in the solution [140]. The results of such calculations for a series of alloys (the variation in surface tension of which on alloying was shown in Figs. 29 and 31) are represented in Figs. 40 and 41.

A characteristic feature of curves expressing the variation of adsorption with composition is the occurrence of maxima differing considerably in magnitude for different surface-active additions. The presence of a maximum on

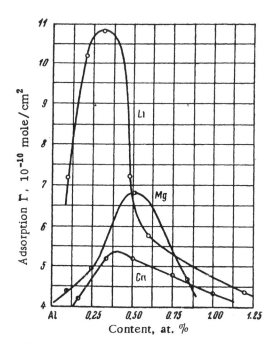

Fig. 40. Variation in adsorption of component introduced into aluminum-based alloys.

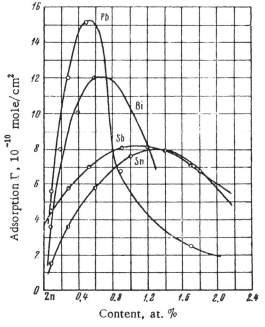

Fig. 41. Variation in adsorption of component introduced into zinc-based alloys.

these curves is due to the determination of adsorption itself as the difference between the concentration of the introduced component in the surface layer C_w and its concentration in the bulk of the solution C, i.e., adsorption $\Gamma = (C_w - C)\sigma$, where σ is the thickness of the adsorbed layer. It follows from this that with increase in the concentration of the second component in the surface layer, the difference in concentration increases at first rapidly, reaches a maximum and then decreases, due to the fact that the volume concentration of the surface-active substance in the surface layer remains almost constant, while the concentration in the principal mass of the solution increases as alloying proceeds.

This is in agreement with the results of work by V. K. Semenchenko [141], who studied the equation expressing the variation of adsorption with concentration in a binary system and found that adsorption must always pass through an extreme value, which is represented by a maximum when a surface-active substance is added to the melt, and a minimum when an inactive substance is added.

By calculation and construction of the adsorption isotherms of a surface-active substance in molten aluminum and zinc, it is possible to determine the volume or area S of a particle of surface-active substance in the surface layer under conditions of complete saturation of the surface by this substance, i.e., under conditions of the formation of a monatomic layer of the surface-active metal on the surface. In doing this, we make use of the well-known equation

$$S = \frac{1}{N \cdot \Gamma_{max}},$$

where N is the Avogadro constant (6.022×10^{23}) and Γ_{max} is the maximum adsorption.

Calculation of the diameter of the particles in the adsorbed film gave the results shown in Table 8.

TABLE 8. Adsorption and Sizes of Particles of a Surface-Active Element in the Surface Layer

Alloys	Γ_{max}, 10^{-10}mole/cm^2	Calculated diameter of particles, A	Atomic diameter of added metal, A	
			K8	K12
Al–Li	10.84	3.9	3.04	3.14
Al–Mg	6.8	4.94	3.10	3.2
Al–Ca	5.22	5.6	3.82	3.94
Zn–Pb	15.17	3.31	3.38	3.48
Zn–Bi	12.0	3.74	–	3.64
Zn–Sb	8.13	4.52	3.12	3.22
Zn–Sn	7.90	4.50	3.06	3.16

The table shows that the calculated value of the diameter of a particle of some metals in the adsorbed layer differs little from the atomic diameter of the same metal obtained from reference data of the dimensions of the crystal lattice, taking the packing of the atoms into account. The agreement between them is closer, the greater the value of maximum adsorption. For some added metals, the calculated value and the value obtained from reference tables differ considerably (for example, for calcium in aluminum and tin in zinc). As a rule, the experimentally determined particle size is somewhat larger than the value of the atomic diameter obtained from reference tables. This may be due to the fact that either the particles in the surface layer of a melt are associated in the form of complexes, or the surface layer is incompletely saturated and is not a monatomic film.

Surface Phenomena in Alloys and Their Utilization in the Production of Castings

Without entering into a discussion of problems connected with the study of surface phenomena in metallurgical processes (oxidation, reduction, desulfurization, electrolysis and so forth) — these problems have been dealt with by B. V. Stark [120], A. M. Samarin [121], S. I. Filippov [139], A. I. Belyaev and E. A. Zhemchuzhina [138], S. M. Baranov [125] and others — reference will be made to work in recent years, which has resulted in appreciable achievements in regard to the improvement of casting processes and to increasing the quality of castings from different alloys.

Still quite recently (in the prewar period), structure modification by surface-active additions was mainly employed in the production of castings of eutectic alloys of silumin type. For other alloys, for example cast iron, the process of structure modification was not widely known or established. Questions of the modification of alloys, in the structure of which primary formations predominate (steel, duralumin, etc.) also had not been developed. Definite progress has currently been made in these fields also. As in the case of the modification of eutectic components, in the refinement of the primary formations, the principal part is played on the one hand by surface phenomena (adsorption of impurities and modifiers on the surface of growing crystals, activation of nometallic inclusions and their conversion into crystallization centers on the adsorption of modifiers on them and so forth), and on the other hand by the increase in number of crystallization centers due to high-melting nuclei.

The value of these processes for the production of quality castings is obvious from the fact, for example, that the extensive application of silumin in industry as one of the principal aluminum casting alloys has been possible only on account of structure modification, i.e., the refinement of the coarse needlelike silicon formations. Structure modification of cast iron by magnesium is gradually acquiring an important position in foundry practice, since it results in a considerable increase in the strength and ductility of cast iron.

Progress in this field could be more substantial if the theory of modification was more fully developed and gave more definite indications regarding the effect of the components of alloys on the structure of the latter on alloying, and the part played by surface-active additions. As will be shown later, there are contradictions in these questions which it is very desirable to remove. This concerns in the first place the fact that a distinction must be made between modification due to a variation in size of the primary macrograin, and modification consisting in the refinement of the structural constituents of secondary formations (eutectice). It is furthermore necessary to define more precisely the concept of modification as relating to a refinement of the dendritic cells of primary separated crystals, provided that the actual size of the dendrites undergoes no change. A consideration of these problems will assist in deciding in which cases the value of the surface tension of a liquid metal is of more particular significance, and in which cases the "nucleation" theory of alloy modification should be used as basis.

Surface Tension and Modification of Alloys of Silumin Type

The most widespread points of departure of the theory of the modification of silumin, developed by the work of Soviet and other investigators, are the concepts regarding the restriction in the linear rate of growth of silicon crystals due to the selective adsorption of the surface-active modifier— sodium— on the faces of these crystals, as well as concepts on the occurrence of supercooling in the crystallization of the eutectic.

TABLE 9. Reduction in Surface Tension of Silumin and Aluminum on the Addition of Sodium

Alloy	σ, d/cm	Remarks
Pure aluminum	849.7	—
The same, with addition of 0.1% Na	695.7	—
Silumin with 10% Si	842.1-831.3	—
The same, with addition of 0.1% Na	619.2	Immediately after addition of sodium
The same	625.2	After 3 min
The same with addition of 0.2% Na	560.9	Immediately after addition of sodium
The same	589.2	After 5 - 10 min

No reliable experimental data are currently available, however, to show the influence of sodium on the value of the surface tension of silumin. Thus, S. V. Sergeev [133] found that sodium in a quantity of 0.1% reduces the surface tension of silumin (and aluminum), sodium being in both cases more surface active than magnesium. The results of the work of A. V. Sergeev confirmed earlier concepts of the part played by sodium in silumin as surface-active component, established by the work of P. A. Rebinder [123], V. K. Semenchenko [33] and others.

Contrary to these results, L. L. Kunin found that sodium is an inactive addition in silumin, increasing to some extent the surface tension of the silumin [79]. In the references cited above, a contradictory opinion is also expressed regarding the function of silicon in this alloy: S. V. Sergeev decided that silicon was inactive in regard to aluminum, while L. L. Kunin considered that silicon was a surface-active component in binary alloys with aluminum.

In view of the directly contradictory conclusions concerning the function of silicon and sodium in Al–Si alloys, we carried out investigations in this field which showed that silicon does not appreciably reduce the surface tension of alloys in a concentration range of from 0.01 to 12% (see Fig. 34). With regard to the influence of sodium on the surface tension of silumin containing 10 - 12% Si, our experiments also definitely showed that the surface tension of silumin is considerably reduced when sodium is added to it. The effect of the sodium addition persists for a comparatively short time, and after the alloy has been allowed to stand for 5 - 15 min, the surface tension of the silumin begins to increase. Sodium also reduces the surface tension of aluminum, which is in agreement with the results of S. V. Sergeev. Table 9 shows the scale of the reduction in surface tension of aluminum and silumin on the addition of sodium.

It will be seen that the reduction in the surface tension of the metal on the addition of sodium was 150 - 250 d/cm or 20 - 30%. S. V. Sergeev found this effect to be only 50 d/cm for aluminum and 80 d/cm for silumin. L. L. Kunin found the increase in the surface tension of silumin on the addition of sodium to be only 20 - 25 d/cm or 4 - 5% of the measured value.

Thus, the function of sodium in the modification of silumin is to reduce the surface tension of the liquid alloy, which points directly to the adsorption of sodium at the liquid–gas interface. It is evident that such adsorption can also take place at the crystal–liquid interface during the solidification of the eutectic, the adsorbent being the silicon crystals, and the substance accumulating round them being possibly sodium atoms or, as considered by A. A. Bochvar and G. M. Kuznetsov [73], molecules of sodium silicide, insoluble in solid silicon.

We also found that besides sodium, additions to silumin which are surface-active are metals such as magnesium, lead, bismuth and others, which reduce the surface tension of the alloy and its base, aluminum.

With regard to the modifying action of lead or bismuth in silumin, it has not yet been possible to obtain definite results indicating the refinement of the structure of the Al+Si eutectic on the addition of lead. One of the problems, the solution of which would fully disclose the nature of the modification process, is to ascertain the temperature and other conditions of casting these alloys which would reveal fully the modifying effect of the above-mentioned surface-active additions. M. V. Mal'tsev [178] remarked that lead does indeed exert a modifying action on alloys of this type. On the basis of our results, it may be quite definitely stated that the modification of the eutectic of industrial silumins by sodium is accompanied by a considerable reduction in surface tension of the liquid alloy, and not by an increase, as stated in some publications. Consequently, the modification mechanism is undoubtedly associated with the adsorption of sodium at the phase boundaries.

Not every surface-active component, however, when introduced into an alloy will produce a change in structure of the eutectic, despite the fact that the effect of the reduction in surface tension on the addition of such a component is much greater than on the addition for example of sodium to silumin. Evidently, for structure modification, in addition to the reduction in surface tension and adsorption at the phase boundaries, it is also necessary that the modifier should assist the supercooling of the eutectic, as pointed out in [73] for the case of eutectic modification.

Fig. 42. Spherulites in rapidly cooled eutectic Al–Si alloys.

Fig. 43. Dendritic and spherulitic crystal forms.

This additional condition for successful modification of the eutectic structure, formed in the normal solidification conditions of castings, becomes a main condition at sufficiently high rates of cooling of the alloy; for example when making silumin castings of small volume, a eutectic with a finely crystalline structure is obtained without the introduction of any special modifiers into the liquid alloy.

Thus, the creation of only one of the conditions for the successful modification of an alloy, i.e., that of attaining a sufficient degree of supercooling of the eutectic on crystallization, can produce a definite result and the eutectic will be modified. As shown by our observations on the structure of silumins of different compositions, the degree of supercooling depends not only on the rate of cooling, but also on the composition of the alloy. This may be judged by the appearance of a spherulitic structure on the surface of small specimens cast in the form of plates on a solid metal slab, since spherulites are formed as the result of strong supercooling [39]. The character of such a silumin structure is shown in Fig. 42.

Fig. 44. Microstructure of rapidly cooled Al—Si eutectic alloy.

The observations showed that other conditions being the same, "incomplete" spherulites are formed in alloys containing up to 6% Si (Fig. 43), completely formed "ideal" spherulites are produced for a content of 8% Si and above. Consequently, the degree of supercooling of an alloy increases as the composition of the alloy approaches the eutectic point (about 12% Si). The primary crystals of aluminum in hypoeutectic and eutectic alloys under these conditions grow from the bottom of the specimen throughout almost its entire height (Fig. 44). In hypereutectic alloys, there is simultaneous formation of primary silicon and aluminum (see Fig. 11b), similar to that observed by A. A. Bochvar in the investigation of the structure of alloys of actual systems. Evidently, in other systems also, the susceptibility of the alloys to supercooling increases as the composition approaches the eutectic point. Thus, we have observed "complete" spherulites in alloys of the Al—Cu system close to the eutectic.

As fundamental conclusion from this section of the work, it follows that the modification of alloys of silumin type is determined by the adsorption of the modifier (sodium or its chemical compounds with silicon) on the boundaries of the growing silicon crystals, whereby the surface tension of the liquid alloys is reduced.

Surface Tension and the Modification of Alloys of Duralumin Type

Surface phenomena and the accompanying variation in structure on crystallization in which modification does not affect the primary crystal formations, but the secondary crystal formations constituting the basis of the alloys (duralumin, brasses, steel and the like), play a large part in the practical production of these alloys. Thus, according to the observations of well-known metallurgists working in the light-alloy field (V. I. Dobatkin, V. A. Livanov and others), ingots having a fine-grain structure are less susceptible to cracking in continuous casting conditions than the same alloys with a columnar, coarse-grain structure. Other technical and also mechanical properties of fine-grain ingots and products made from them are also found to have higher values [68, 78, 80].

Various views are expressed in the literature on the metallurgy of alloys concerning the part played by surface-active alloying additions and the mechanism of alloy modification. By way of example, reference may be made to a number of recent publications indicating a modifying effect in the solidification of steel [79, 124, 125] and light-metal and nonferrous metal alloys [33, 78, 123, 126, 127]. Despite the fact that the mechanism and kinetics of the crystallization of alloys studied by these authors are analogous, structure modification is nevertheless explained differently. Thus, grain refinement of austenitic steel by means of an addition of 0.4 - 0.5% Ca is explained from the standpoint of the adsorption theory, since calcium produces a considerable reduction in surface tension (from 1800 to 1150 erg/cm^2) and, consequently, being adsorbed at the liquid-crystal boundary, it restricts the linear rate of growth of the crystals [79]. In another similar investigation [124] it was found that the addition of small amounts of boron (0.03 - 0.1%) to liquid steel of the austenitic class also reduced the surface tension of the liquid steel, which had a direct effect in refining the grain of the steel and reducing the zone of columnar crystals in castings, due to the adsorption of surface-active boron on the faces of the primary austenitic formations.

The mechanism of the modification of the primary crystals in alloys is elucidated differently in the work carried out in recent years by M. V. Mal'tsev [78, 126]. According to this author's numerous observations, the addition of boron, titanium, zirconium, vanadium and others in an amount of 0.5 - 0.1% to aluminum-based alloys (different categories of duralumin, Al—Mn alloys, etc.) reduces the size of the primary crystals to less than one-tenth or one-hundredth. The columnar or coarsely equiaxial crystals of aluminum solid solution, invariably formed on the solidification of ingots of alloys which do not contain additions of the above-mentioned high-melting metals, are suppressed by the addition of modifiers, and the macrostructure of the ingots becomes uniform and fine-grained. At the same time, the structure of the dendrites themselves is appreciably altered, since the inclusions of secondary crystallization products are finer and are distributed more evenly both on the faces of the dendrites and in their interior, between the branches of the dendrites. The addition of metals exerts the same effect on the structure of ingots of alloys of copper or other metals.

M. V. Mal'tsev explains the fundamental change in structure of alloys by the presence in the modified liquid alloys of many readily available high-melting nuclei, consisting of intermetallic compounds, on which the primary crystals are formed. The occurrence of these high-melting phases has been confirmed by thermal analysis of the solidifying alloys and their separation from the melt by centrifuging the liquid alloys and studying the macrostructure of the centrifuged specimens. It must be stated that this work provides excellent confirmation of the views of A. A. Baikov, who pointed out the importance and function of nuclei and suspended matter, always present in liquid steel and other alloys, and also the views developed by V. I. Danilov and P. D. Dankov on the crystallization processes of liquids in the presence of already-existing, stable crystallization centers.

The results of M. V. Mal'tsev are in good agreement with the previously mentioned results of the work of Japanese investigators [77]. For a number of binary alloys, the composition of which was close to a transition point of the constitutional diagram, they showed that the mechanism of structure refinement is in this case that the primarily separated β-phase crystals (for example in alloys of the Cu—Sn system containing 25% Sn) do not dissolve completely because of the peritectic reaction: β + liquid I → α + liquid I. The rest of these crystals act as additional crystallization centers, producing a refinement of the structure.

In our work on the determination of the surface tension of aluminum when pure and when alloyed with different metals, it was found that the addition of titanium and vanadium had no effect on the value of the surface tension of aluminum and its alloys. Consequently, the structure modification of aluminum alloys (of the type of solid solutions), formed when the alloys solidify in a temperature range, is related to the "nucleation" theory rather than to the adsorption theory.

Industrial alloys of aluminum with manganese or its complex alloys with nickel, copper and silicon, also do not vary their surface tension when alloyed with these components (individually or jointly). The addition of magnesium to these alloys invariably produces a reduction in their surface tension by 110 - 130 d/cm or by 13 - 15%; this is in complete agreement with the results of our measurements shown in the "composition — surface tension" diagrams for alloys of the ternary systems Al—Si—Mg and Al—Zn—Mg (see Figs. 37, 38).

An effective reduction in the surface tension of duralumin is produced by metals such as lead, bismuth, antimony and others which considerably reduce the surface tension of the alloy base, i.e., aluminum (see Fig. 36). Even small additions of these metals in the limits of 0.05 - 0.1% appreciably reduce the surface tension of duralumin, the influence of antimony being the strongest in this case, which may be due to the chemical reaction of antimony with the components of the alloy, with which it forms compounds ($AlSb$, Mg_3Sb_2 and others).

A study of the structure of small ingots of these alloys, cast under normal conditions, indicates that the surface-active additions lead and bismuth are most frequently evenly distributed along the internal boundaries of the dendrites, as shown in Fig. 45. However, the specimens showed no appreciable grain refinement due to the influence of these additions.

Thus, in the modification of aluminum alloys, in the structure of which the primary formations of solid solutions predominate, the principal function of modifiers is fulfilled by additions of inactive metals forming high-melting chemical compounds. The latter produce stable crystallization nuclei in the liquid alloy in the precrystallization period, resulting in a structure refinement of the castings. Of course, the reduction in size of the dendrites is also accompanied by a refinement of their internal structure, with all the resulting favorable consequences in regard to the improvement in the mechanical and other properties of the ingots and articles made from them [68, 78, 126].

In this case, the surface tension of the liquid alloys is not appreciably reduced.

As regards the addition of metals such as antimony, bismuth and others, which considerably reduce the surface tension of duralumin, the part they play evidently amounts to limiting the growth of the secondary dendrite branches (see microstructure in Fig. 45).

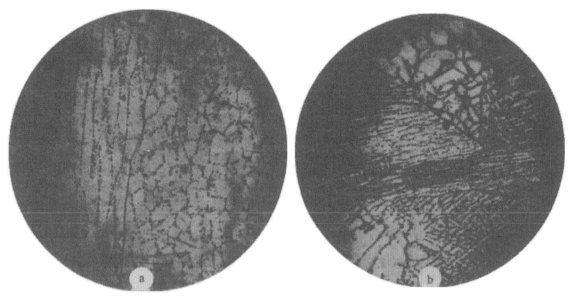

Fig. 45. Uniform distribution of bismuth in aluminum (×90): a) 1% Bi; b) 2% Bi.

Modification of the Industrial Magnesium Alloy ML5

We have also examined experimentally the case of the modification of magnesium alloys, the structure of which consists primarily of crystals of magnesium solid solution. Carbon dioxide was used as modifier. It is known that modification of this alloy may be accomplished in different ways: superheating of the liquid metal in iron crucibles, treatment of the metal with ferric chloride, introduction of small quantities of zirconium and calcium, and treatment with carbonaceous substances [142, 143, 145]. The assumption is that the mechanism of crystal refinement is based on the formation of high-melting crystallization centers, although the nature of the phenomenon has not yet been fully elucidated [145]. The first two methods require high superheating of the liquid alloy (to 900°C), while the last method demands comparatively moderate temperatures. Thus, when marble or magnesite is used as modifier, it is merely necessary to heat the alloy to the decomposition temperature of these salts, i.e., to 780-800°C. The modifying substances — calcium, aluminum carbide and others — are produced as result of the following reactions:

$$CaCO_3 \rightarrow CaO + CO_2$$
$$CaO + Mg \rightarrow Ca + MgO$$
$$CO_2 + 2Mg \rightarrow 2MgO + C$$
$$4Al + 3C \rightarrow Al_4C_3$$

In carrying out the work on the modification of the alloy by carbon dioxide, the object was to reduce the modification temperature as much as possible and diminish the formation of oxides. It is known that the mechanical properties of castings of ML5 alloy, when cast in metal molds, are sometimes found to be very low (lower than those of castings made in sand molds), due to the diminution and almost complete loss of the modifying effect while the metal is standing in the furnace, and also due to its being continuously stirred. Experience with the magnesium alloy shows that the mechanical properties of specimens cast at the commencement of supply of the alloy, i.e., on casting the alloy as just modified, are high; after working for 45 - 60 min, however, a marked reduction in the mechanical properties is observed. It is impossible to effect their improvement by remodification, since the temperature of furnaces does not exceed 700 - 710°C, whereas 780°C is necessary for modification by means of marble or chalk. By using gas modification, the high-temperature stage of carbonate decomposition may be obviated, and carbonaceous inoculation can be produced in the liquid bath at a low temperature.

The experiments were carried out by determining at the outset the extent of the decrease in the properties of the alloy during the standing process (Table 10), and then determining the extent of the improvement in the properties after repeated low-temperature modification (Table 11).

TABLE 10. Decrease in the Mechanical Properties of the Alloy ML5, Modified by Chalk at 790°C, during the Standing Process*

Casting conditions of specimens	Mechanical properties	
	tensile strength kg/mm^2	elongation, %
30 min after commencement of operation of feeding furnace	23.7	9.2
1 hr after commencement of operation of feeding furnace	22.8	7.2
1$\frac{1}{2}$ hr after commencement of operation of feeding furnace	22.5	6.4
2 hr after commencement of operation of feeding furnace	21.3	5.2
After treatment of this melt with carbon dioxide for 7 min	23.5	9.6

* Working temperature of feeding furnace 690° - 705°C. The table gives the mean results of three tests.

TABLE 11. Improvement in the Mechanical Properties of the Alloy in Low-Temperature Modification by Carbon Dioxide (700 - 710°C).

Casting conditions of specimens	Mechanical properties		Duration of modification, min
	tensile strength kg/mm^2	elongation, %	
Before modification (after working for 1.5 - 2 hr	21.3	5.2	—
After treatment of the alloy	23.5	9.6	7
Before modification	22.2	7.8	—
After modification	24.8	8.2	7
Before modification	21.4	6.2	—
After modification	26.2	8.6	10
Before modification	24.8	6.2	—
After modification	26.8	10.4	10

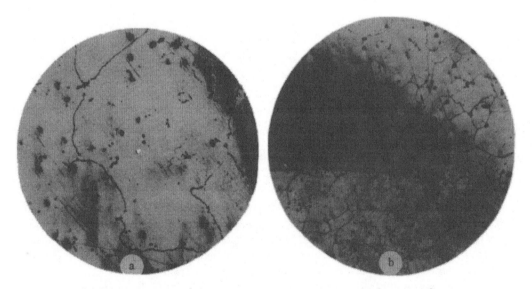

Fig. 46. Microstructure of the magnesium alloy ML5: a) Before modification; b) after modification

Thus, the effect of modifying the metal was to increase the tensile strength by 10 - 20% with an increase in elongation of 1.5 times or more.

Individual experiments carried out to ascertain the influence of the duration of gas injection on the modification temperature show that successful modification is ensured when CO_2 is blown through the melt in 7 - 10 min, a temperature of 700° - 720°C being adequate. The effect of modifying the alloy by means of gas persists during the operation of the melting furnace for 30 - 45 min.

The improvement in the mechanical properties of the alloy after modification is due to the decrease in size of the primary crystals of magnesium solid solution (Fig. 46).

For ascertaining more precisely the modification mechanism of these alloys, their surface tension was determined. The experimental difficulties here consist in the instability of quartz and artificial corundum. Sufficiently reliable and reproducible results were obtained by using iron capillaries, connected to a quartz tube by means of a cement consisting of zinc oxide and water glass.

The experimental results (Table 12) show that aluminum is not surface-active in high magnesium alloys (in contrast to magnesium, which is active in aluminum alloys). This is explained by the relationship of the atomic volumes of these two metals (see Tables 3 and 4).

TABLE 12. Surface Tension of Alloys of the Mg—Al System (Temperature 690° - 720°C, Flux KCl+LiCl)

Alloy composition	Surface tension, d/cm
Mg pure	526.7—521.2
Mg + 1.0% Al	529.3
Mg + 2.0% Al	509.0—516.2
Mg + 2.5% Al	522.4
Mg + 3.0% Al	504.8—510.0
Mg + 5.0% Al	512.6
Mg + 8.0% Al	521.5
Mg + 8.5% Al	521.5
Mg + 10% Al	525.6
Mg + 12.5% Al	520.9

The addition of beryllium and calcium to the alloy ML5 and its treatment with carbon dioxide (Table 13) do not affect the value of the surface tension of the alloy. The structure refinement of these alloys is thus not due to the adsorption mechanism, but depends on the production of crystallization centers in the alloy in the form of chemical compounds, in particular aluminum carbides.

If we compare the results of the modification of the two types of industrial alloys, i.e., the solid-solution type (duralumin, magnesium alloy ML5) and the alloy type with eutectic structure (silumin) it may be said that in the former cases, the principal part is played by high-melting crystallization nuclei, and in the second adsorption and supercooling. Beryllium, titanium and zirconium and duralumin and magnesium alloys do not appreciably reduce the surface ten-

TABLE 13. Surface Tension of the Alloy ML5 on Addition of Beryllium and Calcium and on Modification by Chalk (Temperature 690°-720°C, Flux KCL+LiCl)

Addition, %	Surface tension, d/cm
Alloy ML5	529,1
Beryllium	
0.001	512.2
0.002	512.2
0.003	505.0
0.005	523.5
0.01	523.5
0.05	505.0
0.10	505.0
Calcium	
0.001	509.5
0.005	519.5
0.05	519.5
0.1	523.0
0.3	519.5
0.5	512.2
Chalk (CO₂)* . . .	533—528—541—534

* Temperature 750° - 775°C

sion of the alloys, but the addition of sodium to silumin considerably reduces the surface tension.

It is possible that in other alloys, whose structure is formed in a more complicated way during crystallization than, for example, in simple eutectic alloys of silumin type, modification may be effected by means of both nucleation and adsorption mechanisms. However, this may be, these possibilities of utilizing surface phenomena in the metallurgy of alloys constitute a considerable reserve in regard to improving the service and technological properties of alloys.

Further investigations in this field should be conducted with greater intensity with a view to discovering surface-active additions (modifiers), selectively adsorbed on structural formations of alloys, and to finding additions which assist the mass formation of crystal nuclei, ensuring the orientational growth on them of the structural constituents of the alloys. It is evident that there is a possibility of discovering modifiers of both kinds for the majority of industrial alloys.

Chapter IV

VISCOSITY OF METALS AND ALLOYS

The viscosity of liquid metals is sensitive both to the structural changes occurring in the "pure" liquids with variation in temperature, pressure and other factors, and to the microhomogeneity of commercial metals and alloys.

Measurement of viscosity is one of the methods of the physicochemical analysis of liquid systems, whereby it is possible to disclose the above-mentioned structural features of a liquid.

The concept of viscous flow is based on the correlation of the external forces producing the motion of the liquid, and forces tending to return the displaced layers of liquid to a position of equilibrium as the result of internal friction. Viscosity is, therefore, not infrequently referred to as internal friction.

The fundamental equation expressing the relationship of these forces is represented in the following form:

$$F = \eta S \frac{dv}{dy},$$

where

F is the external force, g·cm/sec^2;

S is the area of the displaced layer, cm^2;

$\frac{dv}{dy}$ is the velocity gradient of displacement of the layers, 1/sec;

η is the coefficient of dynamic viscosity or simply the dynamic viscosity, g/cm·sec.

Thus, the dynamic viscosity of liquids η is defined by the equation

$$\eta = \frac{F}{S \frac{dv}{dy}} \quad \text{g/cm} \cdot \text{sec (poise)}.$$

The kinematic viscosity of liquids is expressed by the equation

$$\nu = \frac{\eta}{d} \quad \text{cm}^2/\text{sec (stoke)},$$

where

d is the density of the liquid,

ν is the coefficient of kinematic viscosity.

The development of the theory of viscosity, like the theory of the surface tension of liquids, is directly connected with the problem of the liquid state of matter and is solved on that basis.

An analysis of the various theories of the viscosity of liquids and a comparison of some of them were made by Ya. I. Frenkel' at a conference on viscosity convened by the Academy of Sciences, USSR [240]. At this conference, theoretical and experimental papers were read by M. P. Volarovich, M. F. Shirokov, A. A. Trapeznikov and other investigators on the viscosity of melts, solutions and oils [240, 241]. Work in this direction has since been reviewed in the monographs of G. M. Panchenkov [242] and E. G. Shvidkovskii [41], and also at a conference convened by the Academy of Sciences, Ukr. SSR at Kiev University [243].

The fundamental premises in the derivation of the viscosity equations are based on concepts of the quasihomogeneous structure of liquid metals. These concepts follow from an exmaination of the mechanism and kinetics of melting and solidification, during which there is a sharp variation in the order of arrangement of the atoms and their mobility, due to a variation in the kinetic energy and the bounding forces between the atoms (see Chapter I).

The concepts on the structure of liquid metals are in good agreement with typical "viscosity – temperature" curves. The systematic decrease in viscosity with increase in temperature is associated here with the "washing away" or erosion of the structure of the liquid metal and the weakening of the forces of interatomic reaction, which is particularly noticeable during variation in temperature close to the melting point of metals.

The results are given below of determinations of the viscosity of some metals and alloys, and the relationship is indicated between the values for the kinematic viscosity of metals and some of their physical properties.

Among the earliest work on the systematic study of the viscosity of liquid metals is that of F. Sauerwald and his co-workers, who employed the method of flow through a capillary. They determined the viscosity of a number of metals and some alloys [244].

TABLE 14. Viscosity of Liquid Metals Near the Melting Point

Metal, °C	Dynamic viscosity, centipoises	Kinetic viscosity, centistokes	Source
Aluminum, 700	–	0.44	
	3.84	1.62	[251]
	2.84–3.55	1.2–1.5	[254]
	1.11	0.47	[249]
	–	0.48	[41]
Bismuth, 300	–	0.16	
	1.662	0.16	[135]
	–	0.18	[41]
Tin, 250	–	0.29	
	1.82	0.26	[135]
	–	0.25 (282°)	[255]
	–	0.28	[41]
Lead, 350	–	0.26	
	2.62	0.25	[135]
	2.44	0.23	[255]
Zinc, 450	–	0.35	
	3.17	0.46	[135]
	–	0.30(500°)	
	–	0.37	[41]
	2.40	0.35	[164]
Antimony, 650	–	0.26	
Antimony, 700	–	0.24	
Antimony, 702	1.29	0.20	[135]

The viscosity of sodium and potassium was studied by Chiong [245], and the viscosity of lithium, rubidium and cesium by Andrade and Dobbs [246] in vacuum using the method of the torsional oscillations of a sphere filled with the metal under investigation.

The viscosity of tin and bismuth was investigated by E. V. Polyak and S. V. Sergeev [247] using the method of damped torsional oscillations of a sphere in the liquid metal. In this method, calculation gives the value of the dynamic viscosity (in poises).

The viscosity of lead, tin and bismuth was also determined by E. G. Shvidkovskii [41] by the method of damped torsional oscillations of a small "cup" filled with the liquid metal. Calculation by the method used by E. G. Shvidkovskii gives values of the kinematic viscosity (in stokes).

The viscosity of aluminum and its alloys was first determined by S. V. Sergeev and E. V. Polyak [248], using the above-mentioned method. Their attempt to determine the viscosity of aluminum by the method of flow through a capillary failed, since despite the inert atmosphere, the artificial carborundum capillary became blocked with oxides. Measurement of the viscosity of aluminum by the method of damped oscillations of a steel sphere in the liquid metal gave results which, compared with later observations, were very high, due to the braking effect exerted on the rod (from which the sphere was suspended) by the surface oxides, and to premature crystallization of the metal on the surface of the metal. Lower values for the viscosity were obtained by E. G. Shvidkovskii [41] and E. Gebhardt [249] using the method of damped oscillations of a cylinder filled with the liquid metal.

The viscosity of aluminum and some other metals has been determined by British investigators using two methods: Yao and Kondic [250] employed a viscometer, operating on the method of the damped oscillations of a cylinder, forcibly immersed in the liquid metal, while Jones and Bartlett [251] determined the viscosity by the method of coaxial cylinders with outer rotating cylinder. This method was subsequently used for investigating the viscosity of copper and alloys of copper with silicon, aluminum and tin [252].

Thus, reference works and original articles give data on the viscosity of liquid metals, obtained by stationary methods (the capillary method and the method of rotating cylinders) and nonstationary methods (torsional oscillation methods). These methods are described in detail in the special literature [253].

Table 14 gives the results of some of the work referred to in the foregoing, as well as results obtained in our laboratory. The values of the density of the liquid metals required for the conversion of the dynamic viscosity values into stokes, were obtained from reference data [135].

The considerable discrepancy in the results for aluminum (2 to 3 times) is evidently due to the presence of oxide films in molten aluminum. Special experiments made by E. G. Shvidkovskii have confirmed this supposition. Thus, the viscosity of aluminum after standing for several hours under slight vacuum was almost twice as high as in the case of aluminum tested in a high vacuum without standing time [41]. The influence of standing time in an oxidizing atmosphere in increasing the viscosity of the metal was also shown in a previously cited work [250]. It was found that the viscosity of zinc superheated to 200°C increased by almost 5 times after standing in the air for 3-4 hr. In the case of slight superheat (50°C), the viscosity of zinc increased insignificantly during this period (approximately from 4 to 6 centipoises). According to these data, aluminum and tin impurities in zinc and titanium in aluminum (up to 0.2%) almost double their viscosity. It should be noted that in this work a high value (about 3 centipoises) was obtained for aluminum.

Figure 47 compares the results of these investigations. It is possible that these results were affected not only by the presence of oxides in the melt but also by the difference in the methods of determining the viscosity. It should be noted that the results of E. G. Shvidkovskii and E. Gebhardt, obtained by the same method (damping oscillations of a cylindrical crucible) are in good agreement (see Table 14).

If the numerous data concerning the viscosity of metals are examined, a striking difference in character is found for the "viscosity – temperature" curves. The most usual form of curve is the one showing a continuous decrease in viscosity with temperature (curve 2 in Fig. 47). Another form of "viscosity – temperature" curve (curves 1 and 4) show a considerable decrease in the influence of the temperature after melting, when the viscosity remains practically constant with increase in temperature.

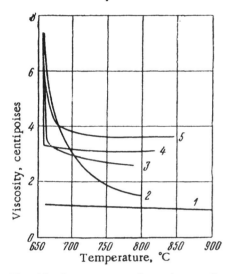

Fig. 47. Temperature dependence of viscosity according to the results of various authors [249]. 1) Gebhardt; 2) Polyak and Sergeev; 3) Yao and Kondic; 4) Santo Matsukava; 5) Jones and Bartlett.

The results we have obtained show a continuous decrease in the viscosity of metals with temperature, which is in agreement with the results of earlier work by F. Sauerwald [244] on copper, antimony and other metals, and also with the results of the latest investigations of E. G. Shvidkovskii (Fig. 48) and E. Gebhardt on tin, lead, magnesium, silver, although for aluminum, the temperature dependence of viscosity is obtained in the form of a straight line (see Fig. 47) in the case of E. Gebhardt. It should be mentioned that the continuous decrease in viscosity with temperature is in agreement with the theory of the viscous flow of liquids proposed by Ya. I. Frenkel'.

The results of most of the above-mentioned work indicate that at a temperature near the melting point of the metals, there is a sharp increase in viscosity, due to variations occurring in the structure of the liquid in the precrystallization period.

The foregoing data on methods of measuring viscosity and its magnitude for different metals relate to the case of laminar flow of the liquid at low velocities, but not to the case of rapid displacement of layers of the liquid, when the formation of eddy motion (turbulent flow) is possible in it.

Construction of Viscometer and Mathematical Formulas

The viscosity of a number of metals and alloys was investigated in our work by means of a vacuum viscometer, the design of which was proposed by E. G. Shvidkovskii [41]; an instrument of similar design is also used for the de-

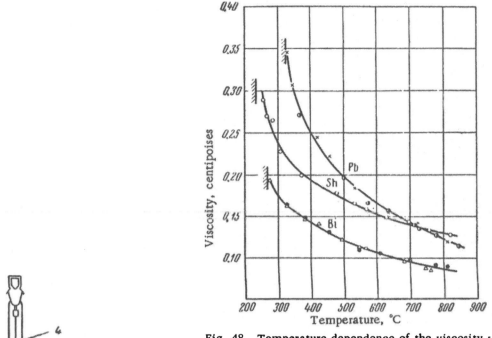

Fig. 48. Temperature dependence of the viscosity of lead, tin and bismuth [41].

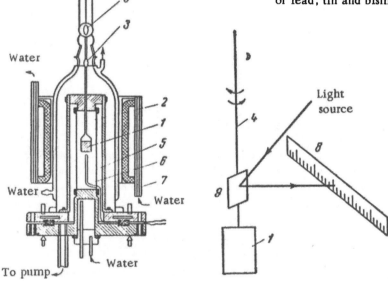

Fig. 49. Construction of viscometer, diagram showing its action and reading of the indications.

termination of the electrical conductivity of liquid metals by the noncontact method [257]. Figure 49 shows the construction of the viscometer. When this viscometer is used, the effect of the surface oxide film is obviously very slight, since the film is connected to the walls of the rotating cylinder only along a line of contact, and mixes with the metal to a lesser degree during the experiment period.

The specimen undergoing test is placed in a cylindrical crucible 1, situated in a quartz suspension 2 connected by a clip 3 to a tungsten filament 4. The tubular sheet molybdenum heater 5 is fed off a current transformer type OSU-40; a rheostat on the high-voltage side enables the temperature to be regulated with sufficient accuracy. The temperature is measured by means of a thermocouple 6, situated below the crucible at the center of the heater; by maintaining the current at a given strength for 5 - 10 min, a uniform temperature is obtained in the space and in the crucible.

The vacuum of 10^{-3} atm, sufficient for our purposes, is produced by means of an oil pump. The apparatus is started up by means of the stator 7 producing a rotating magnetic field, which cooperating with the liquid metal (and graphite crucible) imparts to the system a torsional oscillatory motion about its own axis. This apparatus is enclosed in a quartz cover with double walls between which water is circulated. The stator is switched on for a short time only for "starting" the system, after which damping of the oscillations occurs, due to the internal friction of the liquid metal and the forces of inertia.

The amplitudes of oscillation are measured by means of a reading scale 8, along which travels a moving beam of light reflected from the mirror 9. The period of oscillation is measured by means of a stopwatch. It is usually obtained as the mean of three to five oscillations of the system.

The kinematic viscosity is calculated according to a formula proposed by E. G. Shvidovskii [41]:

$$\nu = \frac{1}{\pi}\left(\frac{K}{MR}\right)^2 \frac{\left(\delta - \delta_0 \dfrac{T}{T_0}\right)^2}{T\sigma^2},\tag{1}$$

where δT and $\delta_0 T_0$ are the logarithmic damping decrements and periods of oscillation of the suspended system with the crucible full (δT) and empty ($\delta_0 T_0$) ($\delta = \frac{1}{n}\ln\frac{A_0}{A_j}$, where A_0 is the value of the initial amplitude; A_j, the value of the final amplitude of oscillation of the system; n is the number of amplitudes measured);

K is the moment of inertia of the suspended system (with the crucible empty);

R is the internal radius of the crucible (radius of the column of metal tested);

M is the mass (weight) of metal tested;

σ is a correction coefficient.

The main difficulties in successfully applying this method are the choice of the tungsten filament with the necessary elastic properties (suitable for obtaining a definite value of the oscillatory period), as well as the dimensions of the crucible and the weight of metal. These values determine the limit of applicability of the principal mathematical formula derived from the equation:

$$\varepsilon = R\sqrt{\frac{2\pi}{\nu T}},\tag{2}$$

where ν is the viscosity, stokes;

T is the oscillation period, sec;

R is the crucible radius, cm.

The mathematical formula is applicable under the following conditions:

$$1.2 > \varepsilon > 10.$$

In testing metals, these being liquids of low viscosity, the criterion ε should not be less than 10. This condition is satisfied for example by the use of a graphite crucible of $D = 15$ mm, $d = 11$ mm ($R = 0.55$ cm) and a tungsten filament 80μ in diameter (previously annealed at $300°C$ for 3 hr); a weight of metal equal to 4 cc then gives a period of oscillation of about 4 sec. Thus, when metals having a viscosity of $0.004 - 0.002$ stoke are tested,

$$\varepsilon = 0.55\sqrt{\frac{2\cdot 3.14}{0.003\cdot 4}} \approx 12.$$

For $\varepsilon > 10$, calculation is performed according to the formula for low-viscosity liquids [41], for $\varepsilon < 1.2$, according to the formula for high-viscosity liquids (not given here). The terms "high viscosity" and "low viscosity" here have a purely arbitrary significance.

In calculating the viscosity, the value of the viscosity (ν^*) without correction coefficient is first determined, and the final value of the viscosity is then determined according to the equation

$$\nu = \frac{\nu^*}{\sigma^2}.\tag{3}$$

The correction coefficient is found from the equation:

$$\sigma = 1 - \frac{3}{2}x - \frac{3}{8}x^2 - a + \frac{4R}{2H}(b - cx). \tag{4}$$

The values of a, b, and c are found from Table 15 as a function of y:

$$y = \frac{2\pi R^2}{\sqrt{T}}; \quad x = \frac{\delta}{2\pi}.$$

The following is an example of the calculation of the viscosity of lead at 485°C in a graphite crucible, in which the height of the metal 2H = 4.5 cm, R = 0.55 cm. The following data were obtained experimentally as the mean of six determinations:

$$\delta = 0.166, \quad T = 4.3 \text{ sec}, \quad M = 46.6 \text{ g}.$$

The moment of inertia of the suspended system without the metal (but with a special disk) is determined from the equation

$$K = \frac{K_1 \cdot T_0^2}{T_1^2 - T_0^2}, \text{ where } K_1 = \frac{mR_1^2}{2}.$$

TABLE 15. Values of the Coefficients a, b, c in (4) Calculated as Functions of y [41]

y	a	b	c	y	a	b	c
100	0.2121	0.0466	0.1115	1800	0.0500	0.1037	0.1688
150	0.1732	0.0587	0.1243	1900	0.0487	0.1042	0.1693
200	0.1500	0.0669	0.1312	2000	0.0474	0.1047	0.1697
250	0.1342	0.0725	0.1366	2100	0.0463	0.1052	0.1701
300	0.1224	0.0765	0.1409	2200	0.0452	0.1056	0.1704
350	0.1130	0.0798	0.1444	2300	0.0442	0.1060	0.1707
400	0.1061	0.0826	0.1472	2400	0.0433	0.1064	0.1710
450	0.1000	0.0850	0.1496	2500	0.0424	0.1067	0.1713
500	0.0947	0.0870	0.1517	2600	0.0416	0.1070	0.1716
600	0.0865	0.0901	0.1552	2700	0.0408	0.1073	0.1718
700	0.0801	0.0926	0.1579	2800	0.0401	0.1076	0.1720
800	0.0750	0.0946	0.1601	2900	0.0394	0.1078	0.1722
900	0.0708	0.0962	0.1618	3000	0.0387	0.1080	0.1724
1000	0.0671	0.0975	0.1631	3100	0.0381	0.1082	0.1725
1100	0.0639	0.0986	0.1642	3200	0.0375	0.1084	0.1726
1200	0.0612	0.0995	0.1651	3300	0.0369	0.1086	0.1727
1300	0.0588	0.1004	0.1659	3400	0.0364	0.1088	0.1728
1400	0.0567	0.1012	0.1666	3500	0.0358	0.1090	0.1729
1500	0.0547	0.1019	0.1672				
1600	0.0530	0.1026	0.1678				
1700	0.0514	0.1032	0.1683				

The weight of the disk m = 14.7 g, radius of the disk R_1 = 2.45 cm, period of oscillation with empty crucible T_0 = 4.356 sec, period of oscillation with empty crucible, together with the disk T_1 = 7.6 sec.

Whence,

$$K_1 = \frac{14.7 \cdot 2.45^2}{2} = 44.1; \quad K = \frac{44.1 \cdot 4.356^2}{7.6^2 - 4.356^2} = 21 \ 4 \text{ g} \cdot \text{cm}^2.$$

Viscosity of metal without correction ($\sigma^2 = 1$)

$$\nu^* = \frac{1}{\pi}\left(\frac{K}{MR}\right)^2 \frac{\left(\delta - \delta_0 \cdot \frac{T}{T_0}\right)^2}{T\sigma^2}.$$

The term $\left(\delta_0 \cdot \frac{T}{T_0}\right)$ of the equation is calculated first from the experiment with the empty crucible:

$$\delta_0 = \frac{1}{n}\ln\frac{A_0}{A_f} = \frac{2.3}{3}\cdot \lg 1.016 = 0.00528.$$

Thus

$$\nu^* = \frac{1}{3.14}\left(\frac{21.4}{46.6\cdot 0.55}\right)^2 \frac{(0.166 - 0.0053)^2}{4.3} = 0.00132 \text{ stoke.}$$

The correction coefficient σ for R = 0.55 cm, 2H = 4.5 cm is calculated from (4) for $y = \frac{2\pi R^2}{\nu^* T} = 330$ and for values of a, b, and c according to Table 15; in this case:

$$x = \frac{\delta}{2\pi} = \frac{0.166}{2\cdot 3.14} = 0.026.$$

Neglecting the third term in (4), in view of its small value, we get:

$$\sigma = 1 - \frac{3}{2}\cdot 0.026 - 0.1177 + \frac{4\cdot 0.55}{4.5}(0.0781 - 0.1424\cdot 0.026) = 0.88.$$

The final value of the viscosity

$$\nu = \frac{\nu^*}{\sigma^2} = \frac{0.00132}{(0.88)^2} = \frac{0.00132}{0.77} = 0.00172 \text{ stoke or } 0.172 \text{ centistoke.}$$

This value is in good agreement with the values found by E. G. Shvidkovskii (0.2) and other workers (0.19 - 0.17 centistoke).

Viscosity of Pure Metals in Relation to Their Physical Characteristics

The data we have obtained on the viscosity of metals are given in Table 14. The dependence of viscosity on temperature for these metals is shown in the diagram of Fig. 50.

Comparison of the values of the kinematic viscosity of pure metals with their various physical properties shows that the viscosity of metals is connected with their atomic volume: viscosity decreases with increase in atomic volume (Fig. 51).

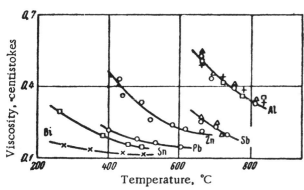

Fig. 50. Dependence of viscosity of metals on temperature.

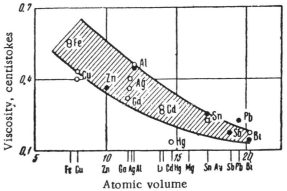

Fig. 51. Relationship between viscosity of pure metals and their atomic volume.

Elementary calculations show that a number approximately characterizing the kinematic viscosity of some metals near the melting point can be obtained from the relationship

$$\nu = K\,\frac{1}{V},$$

where K is a constant, equal to 4 - 5;

   V, the atomic volume of the liquid metal.

   Table 16 shows that the value of the viscosity of a number of metals calculated from this relationship are in good agreement with the experimental values obtained by us and other investigators [41, 164, 247, 249, 256], not merely in order of magnitude but also in absolute value. This is a manifestation of the fact that the resistance to displacement of particles of the liquid metal relative to adjacent particles is inversely proportional to the distance between their centers.

TABLE 16. Comparison of Experimental and Calculated Values of the Kinematic Viscosity of Metals

Metal	Viscosity, centistokes		Metal	Viscosity, centistokes	
	Experimental	Calculated (K=4.5)		Experimental	Calculated (K=4.5)
Copper	0.41	0.55	Mercury	0.144	0.306
Silver	0.38	0.387	Aluminum	0.44-0.47	0.391
Gold	0.30	0.396	Gallium	0.31	0.394
Iron	0.55	0.54	Tin	0.26	0.264
Magnesium	0.63	0.288	Antimony	0.20	0.238
Zinc	0.36-0.38	0.463	Lead...............	0.20	0.23
Cadmium	0.17-0.29	0.32	Bismuth	0.16	0.216

   Exceptions in this connection are magnesium, the experimental value for its viscosity being more than twice the calculated value, and mercury for which, on the contrary, the calculated value is much higher than the experimental value.

   The experimental values of the viscosity of the alkali metals are also 3 to 5 times higher than the calculated values. It is possible that this is connected with their structure (body-centered cubic lattice), and also with their high oxidizability, which cannot always be avoided in experimentation. As known, for aluminum viscosity values 3 to 4 times higher than those found by us and E. G. Shvidkovskii (see Table 14) were formerly obtained, which was evidently due to the effect of oxides.

   Further work on the determination of the viscosity of metals and compounds with different susceptibilities to oxidation and different types of lattice ought to indicate how widely and accurately it will be possible to utilize the hypothesis relating to the connection between the kinematic viscosity of metals and alloys and the value of their atomic volume.

   There is still another physical characteristic which has a direct bearing on the assessment of the viscosity of metals; this is their entropy. Entropy (or increase in entropy) ought to characterize correctly the viscous properties of metals, since it reflects the degree of ordering of the atoms in a system which is undergoing some change [258]. The diagram of Fig. 52 does in fact show that the kinematic viscosity of a metal is higher, the lower is its entropy, i.e., the less intense is the disordering of the atoms on heating.

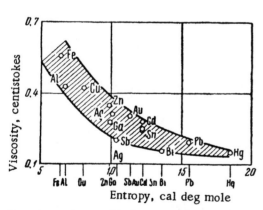

Fig. 52. Relationship between the viscosity of liquid metals and their standard entropy values.

   The foregoing enables us to draw the conclusion that there are two factors which determine the value of kinematic viscosity of liquid metals: 1) the atomic volume as geometric factor and 2) the standard entropy value as energy factor. These factors may be used as objective means of assessing the viscosity of metals.

   Taking into account the practically complete analogy in regard to the function of the two factors in the assessment of the values of kinematic viscosity and surface tension [111], it is to be expected that a direct connection will be found between these properties of liquid metals. The possibility of establishing such a connection is discussed by V. K. Semenchenko [259].

Viscosity of Liquid Alloys and Relationships in the Variations in Viscosity as a Function of Composition and the Form of the Constitutional Diagrams

For binary solutions, N. S. Kurnakov established three classes of "composition − viscosity" diagrams: 1) Continuous curves, differing little from straight lines, connecting the points of the pure components (characteristic of normal nonassociated liquids); 2) curves having a maximum characterizing the presence of chemical compounds in the system; 3) curves having a minimum, corresponding to the formation of the simplest decomposition products of the associated components. N. S. Kurnakov also found a new type of viscosity diagrams with a well-defined maximum, corresponding to a definite chemical compound [260].

N. A. Trifonov [261] analyzed many possible forms of viscosity diagrams for mixtures of two liquids and made a generalization of the factual data. Among the large number of them, only one diagram is mentioned for metallic solutions of the Cu−Sb system (according to F. Sauerwald) characterized by an inflection point on the viscosity isotherms, corresponding to an intermetallic compound. All the other diagrams relate to solutions of organic and inorganic compounds.

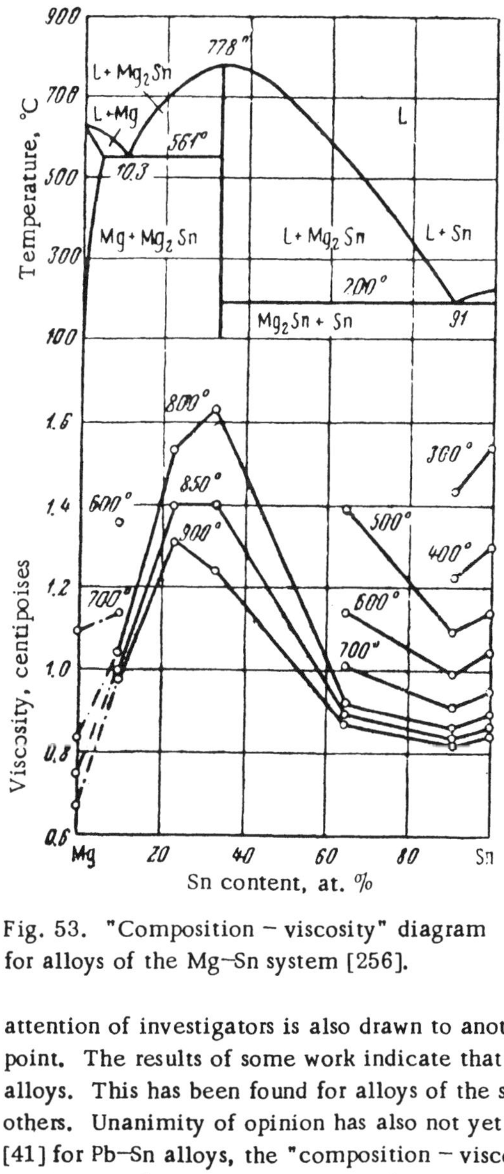

In recent years, E. G. Shvidkovskii [41] and A. Z. Golik and his co-workers [243] have examined questions of the viscosity of metallic alloys in connection with the atomic (molecular) structure of a liquid.

E. Gebhardt et al., [256] showed that in the case of alloys whose components form chemical compounds, the "composition − viscosity" curves have a singular maximum, as will be seen from the example of Mg−Sn alloys (Fig. 53). The same results were obtained for alloys of the Mg−Pb system. In alloys of silver with tin, despite the existence of chemical compounds of the components (Fig. 54), the viscosity curves fail to show such a maximum, evidently because these compounds are formed by peritectic reaction on solidification, while compounds formed by tin and lead with magnesium are precipitated directly from the liquid alloy. In the latter case, the maxima on the viscosity isotherms are smoothed out when the temperature is increased, this increase in temperature resulting in the decomposition of the compounds and the weakening of the bonding forces between the atoms of the alloy components.

When this problem is considered, often no attention is paid to the high value of the temperature coefficient of viscosity of alloy-compounds, and the relatively low value of the coefficient for components and alloys of adjacent composition, or to the different degrees of superheat of the alloys. If the viscosity values of alloys at temperatures of identical superheat above the liquidus curve are plotted, the maxima on the curves are smoothed out, pointing either to the insignificant part played by the "additional" forces of the components in the formation of intermetallic compounds (for example, the compounds Mg_2Sn and Mg_2Pb), or to the practical complete decomposition of these compounds on melting.

Fig. 53. "Composition − viscosity" diagram for alloys of the Mg−Sn system [256].

When the variations in viscosity of alloys are analyzed, the attention of investigators is also drawn to another critical point on the constitutional diagrams, i.e., the eutectic point. The results of some work indicate that alloys of eutectic composition have the lowest viscosity of adjacent alloys. This has been found for alloys of the systems: Zn−Sn [250], Pb−Sn [262], Al−Si [248, 251], Mg−Pb [256] and others. Unanimity of opinion has also not yet been reached, however, in regard to this question; thus, in one case [41] for Pb−Sn alloys, the "composition − viscosity" curves showed a slight maximum for the eutectic instead of a minimum, while in others [247] and [254], neither an increase nor a decrease in viscosity was found.

Nor is there any certainty also with regard to character of the viscosity curve for alloys of composition corresponding to the region of limited solid solutions. Thus, an increase in viscosity was found for Al−Si, Al−Mg and

Al—Cu alloys in the region of solid solutions on an aluminum basis [251]. At the same time, in other work [252], the same authors, for copper-based alloys, found that each of the regions of α, β and γ solid solutions corresponds to a separate portion of the "composition – viscosity" curve, the viscosity of these alloys being higher than the viscosity

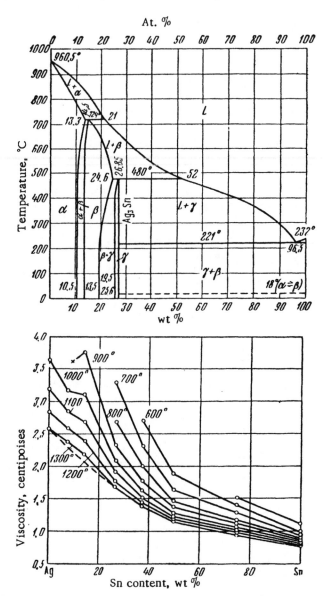

Fig. 54. "Composition – viscosity" diagram for alloys of the Ag—Sn system [256].

of copper. On the other hand, for Al—Sn alloys [164] no increase in viscosity for solid solutions was observed, while for solutions of lead in tin, a reduction in viscosity compared with lead was found [254]. The same was also found in the case of Ag—Sn alloys for solid solutions on a silver basis [256].

Thus, this brief review of the viscosity of alloys shows that the problem of discovering the relationship governing the variation of this interesting property of liquid metallic solutions is far from being solved.

We have investigated the viscosity of a number of aluminum-based alloys by the above-described method of torsional oscillations of a cylindrical crucible filled with liquid alloy. The alloys of aluminum with copper, silicon, and iron were examined in open graphite crucibles, while the alloys containing zinc were used in artificial corundum crucibles sealed in quartz ampoules.

The most characteristic viscosity curves were obtained for alloys of the Al—Cu system (Fig. 55). In this case, we can definitely state that the kinematic viscosity of alloys situated close to the eutectic point (27 - 30% Cu) has a minimum value; it is less than half that of aluminum or that of the chemical compound Al$_2$Cu (54.1% Cu).

For Al–Si alloys (silumins), the same regularity is observed; the viscosity isotherms show that alloys of eutectic composition (11.7% Si) have a lower viscosity than the adjacent alloys (Fig. 56). On the viscosity curves of these alloys, heated by the same amount above the liquidus, however, the minimum is less pronounced than on the isotherms; nevertheless, we can definitely say that the low alloys and hypereutectic alloys have a higher viscosity than the eutectic alloys.

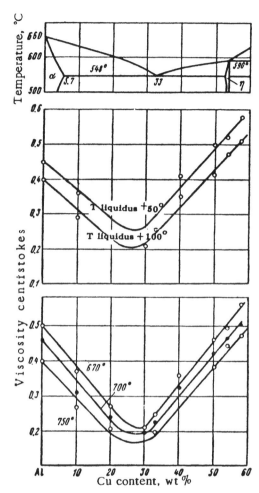

Fig. 55. "Composition – viscosity" diagram for Al–Cu alloys.

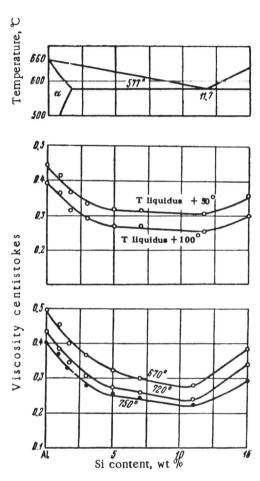

Fig. 56. "Composition – viscosity" diagram for Al–Si alloys.

Similar results were obtained for alloys of the Al–Ge system (Fig. 57), corresponding to the analogy in the equilibrium diagrams of these systems.

The influence of iron on the viscosity of aluminum is shown in Fig. 58; here again, it is found that alloys of eutectic composition have minimum viscosity.

Alloys of tin with zinc and aluminum have a viscosity which gradually decreases with increase in the alloying component (Figs. 59 and 60). It must here be pointed out that alloys situated in the region of the solid solution of zinc in aluminum have a lower viscosity than aluminum (and zinc). This is in contradiction to the above-mentioned investigations on aluminum alloys, according to which the viscosity of the alloys (solid solutions) is higher than the viscosity of the solvent (aluminum).

It is obvious that this problem calls for further investigation; it is possible that not all solvent metals will experience an increase in their viscosity in the liquid state as the result of the entry of other metals into their lattice, i.e., not all solid solutions will have, after melting, a structure different from the structure of the liquid metal free from impurities and alloying additions.

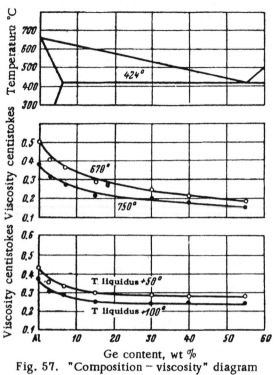

Fig. 57. "Composition — viscosity" diagram for Al—Ge alloys.

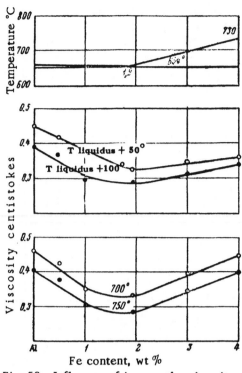

Fig. 58. Influence of iron on the viscosity of aluminum.

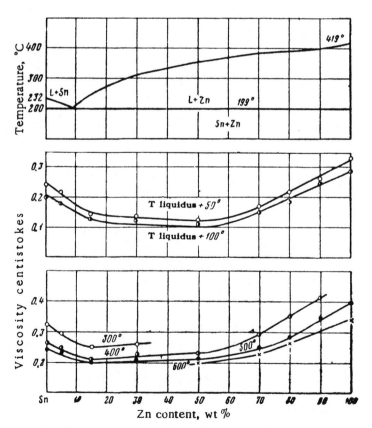

Fig. 59. "Composition — viscosity" diagram for Zn—Sn alloys.

With regard to the viscosity of eutectic alloys, the reduction in their viscosity compared with other alloys is evidently due to the fact that in such alloys there occurs a predominance of the bonding forces between like atoms, but not between unlike atoms, i.e., a comparatively weak interatomic reaction. It is also possible that the atomic volume of the liquid alloy of eutectic composition is larger than the atomic volume of adjacent alloys, due to the low temperature of the eutectic near the melting point. This is in agreement with the fact that liquid metals of

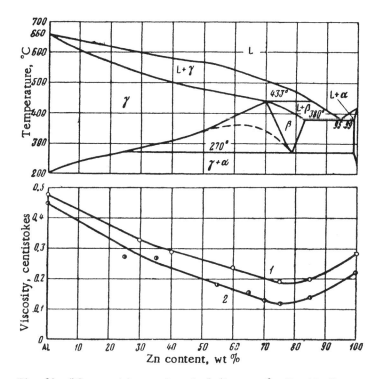

Fig. 60. "Composition — viscosity" diagram for Zn—Al alloys
1) Viscosity at 700°C according to E. Gebhardt; 2) the same
according to the results of the author's laboratory.

high atomic volume have a comparatively low kinematic viscosity. The other fact, entropy, acts in the same direction (decrease in viscosity), since the value of the melting entropies of eutectics (ratio of melting heat to melting temperature) will always be higher than that of the components and adjacent alloys, due to the melting point minimum of the eutectic.

Thus, in an assessment of the viscosity of alloys, the position of the alloys in the constitutional diagram must be taken into account, and it is also necessary to make use of the geometric and energy factors (atomic volume and entropy), which determine the magnitude of the kinematic viscosity.

Chapter V

THE FLUIDITY OF METALS AND ALLOYS

The filling of foundry molds with liquid metal, its flow through the channels and cavities of the molds, is a complex physicochemical and hydromechanical process. To be able to control this process, we require to have a knowledge of the properties of the metals and various alloys in the liquid state, such as density, viscosity and velocity of flow, surface tension, oxidizability, properties of the oxides, wetting capacity and so forth.

For equality of the temperatures of the liquid metal and the walls of gates and mold, the flow of liquid metal will obey the laws of hydraulics of ordinary liquids, provided we ignore questions concerning the chemical reaction of the metal with the atmosphere and the walls of the mold. If the metal is heated above the crystallization temperature, the possibility of filling the mold depends solely on a slight reserve of hydrostatic head. In given cross sections of a channel, this factor will be determined by the rate of flow of the metal, its velocity of flow and also by the nature of the flow of liquid metal, i.e., whether it is turbulent or laminar.

Not infrequently, the equations of hydraulics are used for determining the rate of flow of metal in the case of free flow from an orifice (from a ladle), for calculating the time required to fill a capacity and also for assessing the character of the motion of a metal in channels. The data required for these calculations, i.e., the value of the dynamic or kinematic viscosity, critical velocity of flow, Reynolds number, etc., are assumed to be those for a metal in a definitely fluid state. In the channels of molds, however, the metal is flowing in conditions of considerable inequality between the temperature of the metal and that of the walls of the mold. It is thus inevitable that there will be a change in the condition of the liquid metal; it will contain crystal nuclei and crystals which have formed; there will, therefore, be a considerable change in the properties of the liquid metal. The gradual accumulation of solidification products in the stream of liquid metal with increase in the duration of flow and on contact with the relatively cold walls of the mold results in a continuous variation in the properties of the liquid metal, such as viscosity and fluidity, and the liquid alloys themselves will be transformed into liquid-solid mixtures. The equations of hydraulics can therefore be used for the processes of flow and the filling of the cavities of ordinary foundry molds only with considerable limitations, while the use of such equations is quite impossible in the case of castings of complicated contour, large surfaces and thin sections.

All these factors have resulted in the development of concepts of fluidity, as being the ability (property) of metals and alloys to fill molds. With the help of these concepts, it is possible to make an assessment of the flow of metal in regard to the cold channels of the mold. The value of the fluidity of any metal and alloy, as a physical characteristic, can be ascertained for absolutely definite conditions of temperature and rate of pouring the metal into a mold with channels of definite cross section, just as when ascertaining such physical characteristics of a metal as ductility or creep. The absence of any appreciable reaction of the metal with the material of the mold or apparatus in which the fluidity is measured is an essential condition in this determination.

Investigations in the last few years have shown that the value of fluidity of a metal may indicate the amount and character of nonmetallic inclusions in steels [121]. In addition, there is a direct connection between fluidity criteria and the mechanical (particularly the plastic) properties of steel [149] and the quality of castings [150].

The above-mentioned complexity of the phenomena observed when a metal flows into a mold have given rise to a variety of methods of determining fluidity. The results of measurements of fluidity obtained by means of the various methods do not differ in accuracy and mainly provide an idea of the variation in fluidity of the various alloys (cast iron, steel, bronze, light alloys) when the technical factors are varied.

The character of the variation in any technical property of alloys, which is of significance in the production of castings, is studied most expediently as a function of the composition in association with the constitutional diagrams of the alloys, and this provides reliable physicochemical bases for foundry practice. The early investigations of A. Portevin and P. Bastien on questions of the fluidity of alloys showed that the variation in fluidity of alloys as a function of the variation in composition obeys certain lays [151].

Prominent among the investigations of the last few years is the work of Academician A. A. Bochvar in ascertaining the general laws of variation of the casting properties of alloys, such as fluidity, shrinkage [26] and the mobility of the metallic liquid between the growing crystals [74], as well as work on the properties of alloys in the liquid-solid state [152]. Numerous investigations have been carried out on the fluidity of ferrous alloys by Yu. A. Nekhendzi and A. M. Samarin [80, 129, 149, 150], B. B. Gulyaev [55] and others.

The following is a brief summary of the substance of the views of these authors on the nature of the fluidity of metals and alloys of different compositions.

A. A. Bochvar considers that the ability of metals and alloys to fill a mold (fluidity) is diminished and disappears as crystallization advances. Since the character of the crystallization of alloys is influenced by their composition, fluidity will depend on the position of the alloy in the constitutional diagram. Thus, for equal superheat of the metal above the liquidus temperature in the case of a continuous series of solid solutions, the pure components will have maximum fluidity; minimum fluidity will be possessed by those alloys of this series for which dendritic segregation will occur to the maximum extent. In the case of alloys of eutectic type, the pure components and the alloy of eutectic composition will possess maximum fluidity. The relatively high fluidity of pure metals and eutectics is due to the fact that during their solidification, crystals of constant composition are formed and grow from the surface of the casting with a continuous front, and the liquid melt (also of constant composition) is, therefore, able to move freely inside the casting.

In alloys solidifying in a temperature range, especially in the case of solid solutions, crystallization proceeds with the formation of threadlike crystals. They are unable to become thicker, because of the variation in composition of the liquid, and they penetrate far into the body of the casting in the form of thin branching dendrites, resulting in a considerable reduction in fluidity.

Yu. A. Nekhendzi and A. M. Samarin also consider that alloys of the solid-solution type (steel) cease to flow in the crystallization range at a temperature of zero fluidity. They propose the following formula for determining fluidity, which takes into account the physical and thermal properties of metals and alloys:

$$\lambda = Ad\,\frac{c(t_1 - t_0) + L}{t_{met} - t_{mold}},$$

where A is a constant, depending on the physicochemical and technical properties of the given metal and mold;

d and c are density and specific heat;

L is the heat of crystallization of the quantity of solid phase on which zero fluidity depended;

t_{met} and t_{mold} characterize the heat exchange between metal and mold during the time the metal was in the liquid state on cooling from t_1 to t_0.

In the opinion of B. B. Gulyaev, flow of metal ceases the moment the end surface of the stream forms a solid plug, the strength of which is sufficient to withstand the hydrostatic head. In accordance with this fundamental prerequisite, he proposed a formula for determining the fluidity of steel, based on the laws of hydraulics of ordinary fluids. The length of the spiral should be equal to the product of the velocity of flow by time, expressed in his formula:

$$\lambda = \sqrt{\frac{2g \cdot H}{1 + \Sigma i}}\,(AR^2 \ln B \cdot t + CHR),$$

where g is the acceleration due to gravity (9.81 m/sec^2);

H is the hydrostatic head;

Σi is the total loss of head;

R is the radius of channel;

t is the superheat.

A, B, C are constant magnitudes, related to the thermophysical properties of metal and mold, the strength of the surface film, surface tension and individual features of the melt.

In considering the question of the fluidity of metals and alloys of different compositions, many factors have thus to be taking into account. As shown, however, by the numerous data obtained by N. N. Kurnakov, N. N. Sirota, and M. Ya. Troneva in a study of the fluidity of Fe–Si, Fe–P and other alloys, a decisive influence is exerted on fluidity by the character of crystallization, which in its turn is determined by the position of the alloys in the constitutional diagram [153-155]. The results of these investigations are in full agreement with the deductions of A. A. Bochvar.

On this basis, liquid alloys flowing along the channels and in cavities of a mold may be regarded rather as liquid-solid substances with a varying quantity of liquid and solid phases, and not as ordinary fluids. When the problem is regarded in this light, when the fundamental conditions determining the fluidity of alloys are thermal and structural factors, the function of surface tension and viscosity in the assessment of fluidity, as pointed out by some investigators [156-158 and others], cannot be essential. It is better rather to speak of the influence of the increasing quantity of solid phase in the liquid-solid mixture which, as shown by E. G. Shvidkovskii [41], increases by factors of tens and hundreds. The action of surface oxides films and oxides, suspended in the liquid metal and reducing its fluidity, may also be taken into account.

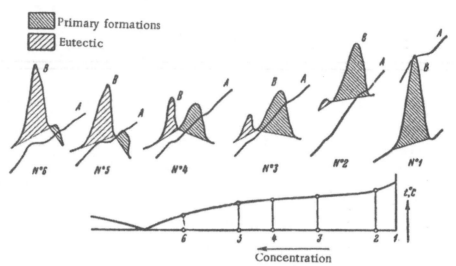

Fig. 61. Variation of the quantity of heat of crystallization of alloys as a function of their composition.

Of the papers published abroad in the last few years on investigations to discover the relationships governing the variation in fluidity of alloys, reference must be made to a paper by V. Kondic [156], describing an investigation of the fluidity of alloys of Al-Si and Pb-Sn systems. It was found that under suitable temperature conditions of casting, the fluidity of aluminum alloys drops sharply on the addition of 1 - 1.5% Si to the aluminum, then remains practically constant at 2 - 12% and increases rapidly at 15% Si. The drop in fluidity is ascribed by the author to the increase in viscosity of the liquid alloys when the aluminum is slightly alloyed, while the increase in fluidity in the region of hypereutectic alloys is ascribed to the greater heat of crystallization of the alloy.

For alloys of the Pb-Sn system, a "composition − fluidity" curve was found, with descending branches close to the components and a maximum at the eutectic point. Similar curves were obtained for eutectic type alloys in the earlier work previously mentioned [26, 151].

In his next paper, V. Kondic [159] speaks of fundamental factors which have a favorable influence on fluidity, and considers that a narrow solidification range, low surface tension and high heat content (determined by the casting temperature) increase the fluidity and vice versa. He considers the influence of surface tension and heat content on fluidity to be incontestable, while the influence of solidification range is debatable. As will be shown, the influence of solidification range is quite incontestable, while the influence of the surface tension range may, in fact, be subject to doubt.

The "composition − fluidity" curves of alloys of the Al-Mg system, published in Metals Handbooks for 1943 and 1948, are also open to criticism; according to these curves, fluidity increases when these metals are alloyed with one another [160]. On the basis of the results of the work discussed in the foregoing on the regular variation of the fluidity of alloys of eutectic type, the fluidity of Al-Mg alloys ought to diminish when these metals are alloyed, a point which has been confirmed in one of our investigations.

Among other investigations in this direction, reference should be made to work by J. Czikel and T. Grossman [161] who have proposed formulas for determining the distance through which a liquid metal travels, the coefficient of resistance and the time elapsing until flow ceases.

Reverting to the formulas examined above for determining the fluidity of alloys, it should be pointed out that one of the basic components in the Nekhendzi − Samarin formula, the quantity of liberated heat of crystallization, is not an absolutely definite magnitude. The authors of the formula assume that zero fluidity of an alloy sets in when about 20% of the heat of crystallization has been abstracted from the liquid metal.

An analysis of the crystallization process of alloys of various compositions shows that not every alloy reaches zero fluidity after losing 20% of the heat of crystallization. This is shown directly by an examination of the results obtained by the author in the thermal analysis of alloys of eutectic type, in which the differential cooling curves shown in Fig. 61 were recorded. The thermal curves A denote the temperature of the primary and secondary formations, while the differential curves B enable the quantity of heat separated in primary and secondary crystallization to be determined (from the area circumscribed by the differential curve). Figure 61 shows that on alloying the heat of crystallization of the primary formations decreases as we move away from the pure component, while that of the secondary formations increases.

It is evident that the assertion made by Yu. A. Nekhendzi and A. M. Samarin to the effect that zero fluidity sets in after removal of 20% of the heat of crystallization is true only of a limited number of alloys. Thus, for example, for alloy No. 5, the loss in heat of crystallization of the primary formations does not reduce the fluidity to zero, since the quantity of primary formations is not large, while the bulk of the alloy is in the liquid condition. The subsequent secondary crystallization will be accompanied by the liberation of its heat of crystallization and will, thereby, help to keep a certain part of the alloy in the liquid state. For alloy No. 2 or No. 3, on the contrary, the removal of the same 20% of heat of crystallization of the primary formations, provided crystallization occurs over a wide temperature range, will certainly produce a reduction in fluidity and may result in zero fluidity. In this case, the heat of crystallization of the secondary formations will be unavailing, since a considerable proportion of the mass of the alloy is already crystallizing in the form of thin, ramified dendrites, preventing the flow of the remaining liquid portion of the alloy.

In the above-mentioned paper by A. A. Bochvar and V. V. Kuzina [74], it was shown that in alloys of the Al–Cu system, during the crystallization of pure aluminum, the eutectic Al+Al$_2$Cu (33% Cu) and the compound Al$_2$Cu (54.1% Cu), the liquid alloy is capable of flowing (flowing out) after 0.9 of the total time of crystallization. In the alloy of aluminum with 5% Cu, crystallizing in a wide temperature range, the flow of liquid stops after 0.25 of the time of total crystallization, when the quantity of liquid is still very large and exceeds the quantity of crystals formed. The quantity of heat of crystallization will be correspondingly small, while in the crystallization of aluminum and the 70 - 80% eutectic, the heat of the crystallizing mass of the alloy forms a considerable proportion of the heat which keeps the remainder of the metal in the fluid mobile condition. To obtain the same fluidity in alloys of different compositions, therefore, a different degree of superheat is necessary; it should be higher for alloys crystallizing in a wide temperature range.

In studying this question, the author plotted solidification curves for Al–Si and Al–Mg alloys. Comparison of these curves with those of Fig. 61 and with the results of fluidity determinations of these alloys shows directly that the heat of crystallization plays a fundamental part in the determination of fluidity.

It must be assumed that the thermal conductivity of metals and alloys will also play a definite part, since it determines the ability of a metal to give off the heat stored up in the liquid metal. This follows for example from the fact that the thermal conductivity of many metals in the solid state is diminished several times under the influence of a small amount of alloying elements entering into solid solution, as will be appreciated from the form of the curves in Fig. 62. It is obvious that such a considerable reduction in thermal conductivity will affect the rate of transfer of heat from crystallizing alloys, which in its turn exerts an influence on their fluidity. This is confirmed by the data of the above-mentioned paper [74]. Thus, the time of total crystallization of equal quantities of pure aluminum and the eutectic alloy of aluminum with 33% Cu in identical thermal conditions of cooling is 4.5 - 5 min, while for an alloy of aluminum with 5 and 12% Cu, it is 8 min. It is evident that such a pronounced increase in the time of total crystallization is due not only to the reduction in the solidification temperature of these alloys, compared with aluminum, but also to the reduction in their thermal conductivity. The increase in the time of total

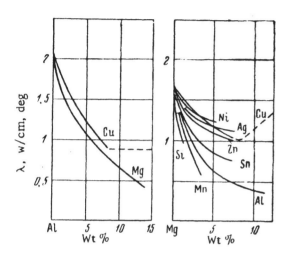

Fig. 62. Variation in thermal conductivity of alloys of aluminum and magnesium as a function of composition.

crystallization of these alloys by no means, however, increases the fluidity; on the contrary, for equal superheat, it is much lower than for aluminum or the alloy of eutectic composition.

It is evident that we are concerned here with a complex action of two thermal characteristics: the relative decrease in the quantity of heat of primary crystallization compared with the pure metal produces a pronounced reduction in fluidity, while the reduction in thermal conductivity, by prolonging the time in which the alloy is in the liquid-solid state, increases fluidity to a very slight degree only. Since the first factor acts directly, while the second leads to the development of an unfavorable structure (dendritic growth), the overall effect of alloying metals with comparatively small additions appears as a considerable reduction in fluidity.

Contemporary concepts of the nature of the fluidity of metals and alloys are thus based on the laws of crystallization and on the variation of thermal physical properties of liquid-solid metal mixtures, such as those which may be formed by alloys during the filling of the mold.

The combination of the thermal physical properties which determine the fluidity of alloys should include the heat of crystallization of the primary and secondary formations of the given alloy composition and their thermal conductivity. It must not be expected that this will exhaust the analysis of all the causes underlying the variation in the ability of metals when alloyed to fill foundry molds. There are a number of other factors (wettability of the walls of molds by the liquid melt, the oxidizability of the latter and the nature of the resulting oxides, their stability on the surface of the metal and their tendency to dissolve in it, and so forth), which will not be discussed here.

Fluidity of Pure Metals

We determined the fluidity of metals by the method most widely used, the spiral method, in the apparatus shown in Fig. 63. This apparatus comprises a cast-iron plate 1, in which is cut a spiral groove 25 mm² in cross section and 2.5 m in total length. The cast-iron spiral mold is heated above and below by two furnaces 2, provided with thermocouples 3. The metal is melted in an electric crucible 4, having in its bottom an orifice, closed by a metal stopper 5.

An apparatus of similar construction is described by A. A. Seminov and was used by him for determining the fluidity of typographical metals [162]. The construction of the apparatus is such as to ensure wide limits in the variation of the heating temperatures of mold and metal, and to maintain constant metal pouring conditions and consequently good reproducibility of the results. The hydrostatic head is always the same, since the volume of metal flowing simultaneously into the mold is small compared with its total reserve in the crucible. The inner walls of the crucible, stopper and surface of the spiral are treated with a special permanent mold coating to prevent any reaction between the liquid metal and the material of the crucible and mold.

The choice of the temperature to which the metal was heated before being poured, and the temperature to which the mold was heated was governed by the desire to have thermal conditions such that it would be possible to compare the fluidity of different metals and examine the values in the light of the thermophysical properties of the metals. Approximation to thermal similarity in carrying out the experiments was obtained by arranging for the temperature of the poured metal to exceed its melting point by a definite percentage of the absolute temperature of the melting point. The temperature of the mold was fixed below the solidification point by an amount which also formed a definite percentage of the absolute temperature on the solidus curve.

In determining the fluidity of pure metals, it was found to be most expedient to take the pouring temperature of the metal as equal to the melting point plus 5% of the absolute temperature of the melting point. The heating temperature of the mold was taken as equal to the melting point minus 15% of the melting point of the metal in degrees absolute. Thus, for zinc, this was:

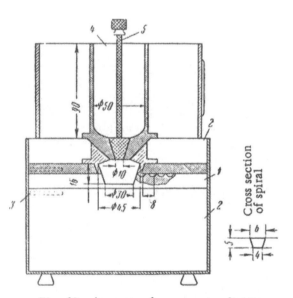

Fig. 63. Apparatus for measuring fluidity.

$$t_{met} = 419 + 0.05\,(419 + 273) = 453°C,$$
$$t_{mold} = 419 - 0.15\,(419 + 273) = 316°C.$$

The experimental conditions and the results (mean values of three determinations) are given in Table 17.

TABLE 17. Comparative Fluidity of Metals

Metal	Pure metal, %	Pouring temperature, °C	Mold temperature, °C	Fluidity (length of spiral), cm
Zinc	99.99	453	316	110
Aluminum	99.75	706	520	95
Tin	99.9	257	156	80
Lead	99.9	357	237	68
Cadmium	99.03	350	232	67
Bismuth	96.8	298	189	60
Antimony	99.0	670	510	50

The data given in the table show that of the pure metals tested, zinc has the highest fluidity and antimony the lowest. It should be mentioned that some of the metals investigated, especially bismuth, cadmium, and antimony, were not quite pure; the fluidity values found for them may therefore be lower than the true values.

It follows from the formulas given in the foregoing for determining the fluidity of metals that this property depends directly on the quantity of heat stored up in the liquid metal. The heat content of the liquid metal up to its total crystallization is composed of the superheat above the melting point Q_1 and the heat of crystallization Q_2.

TABLE 18. Fludity and Heat of Crystallization of Metals

Metal	Fluidity (length of spiral), cm	Pouring temperature, °C	Weight of spiral, g	Specific heat in liquid state, cal/deg	Superheat Q_1, cal	Heat of crystallization (specific) cal/g	Heat of crystallization Q_2 of spiral, cal	$\dfrac{Q_2}{Q_1 + Q_2}$,
Zinc........	110	453	190	0.122	788	25.6	4864	86.1
Aluminum ...	95	706	61	0.262	735	87	5307	86.6
Tin	80	257	144	0.056	201.5	13.45	1937	90.6
Lead	68	357	187	0.0323	181	5.6	1047	85.3

The data of Table 18 show the relationship between the value of fluidity and the heat content of the cast spiral. For the same temperature conditions of pouring, the ability of a given metal to fill a mold, i.e., its fluidity, is determined by the quantity of heat given to the mold by the metal up to the moment of its solidification.

Since the heat of crystallization Q_2 is 85 - 90% of the total heat, it must be assumed that it is just this factor which is decisive in estimating the value of fluidity. External factors, such as superheat of the liquid metal, value of the pressure head, and temperature of the mold, will of course exert a substantial influence on the value of fluidity, increasing or decreasing the actual ability of the metal to fill a mold. For a definite character of crystallization, this ability is mainly determined by the heat of crystallization of the metal.

In confirmation of the fluidity data of Tables 17 and 18, the experiments were made in different thermal conditions, i.e., for a different degree of superheat of the liquid metals and for a lower mold temperature. As shown by the results given in Table 19 and Fig. 64, the fluidity of zinc for any pouring conditions is higher than in the case of tin and lead, which confirms the decisive significance of the heat of crystallization in estimating the fluidity of metals.

The curves in Fig. 64 also show the fluidity of zinc and lead increase considerably only up to a definite superheat, after which is increases much more slowly. A similar effect of superheat on the fluidity of steel is described Yu. A. Nekhendzi [80]. If the rate of cooling of the metal during pouring is varied by reducing the mold temperature, there is then obtained, as will be seen from Table 20, merely a proportionate reduction in the length of the

TABLE 19. Influence of Superheat Temperature on the Fluidity of Metals

Metal	Superheat temperature above liquid-us, °C	Length of spiral, g	Weight of spiral, g	Heat of superheat Q_1, cal	Heat of crys-tallization Q_2, cal	$\dfrac{Q_1}{Q_1 + Q_2} \cdot \%$
Zinc	34	110	190	788	4864	13.9
	41	122	210.45	1052.7	5387.5	16.34
	51	129	222.5	1384.4	5696	19.55
	81	132	227.7	2249.7	5829	27.85
Tin	25	80	144	201.5	1937	9.4
	50	93	167.4	468.7	2251.5	17.23
	98	117	210.5	1155.7	2832.25	28.9
Lead	30	68	187	181	1047	14.7
	50	74	203.5	328.65	1139.6	22.38
	93	83	228.25	685.6	1278.2	38.87
	193	93	255.75	1534.3	1432.2	52.67

spiral, but the metals still arrange themselves in the same order of fluidity, since the thermal conditions prevailing during the pouring of the metals remained the same.

In examining the data obtained in regard to the relationship between the fluidity of metals and the heat of crystallization, it should be noted that the specific heat of crystallization of aluminum is more than three times the heat of crystallization of zinc, while the fluidity of the latter is higher than that of aluminum (see Table 17). The obvious explanation of this is that the thermal con-

TABLE 20. Influence of Mold Temperature on Fluidity of Metals

Metal	Temperature of metal, °C	Mold tempera-ture, °C	Fluidity (length of spiral), cm
Zinc	453	316	110
	453	176	65
Aluminum	706	520	95
	706	340	52.5
Tin	257	156	80
	257	78	42
Lead	357	237	68
	357	117	37

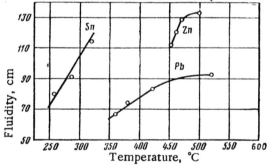

Fig. 64. Dependence of fluidity of pure metals on superheat temperature.

ductivity of aluminum near the melting point is 1.5 - 1.6 times that of zinc, resulting in a relative reduction in the fluidity of aluminum.

Fluidity, Surface Tension, Viscosity, and Velocity of Flow of Pure Metals

It has already been mentioned that in a number of investigations made by other authors, considerable significance was attached to the surface tension and viscosity of metals in estimating their fluidity [156 - 159].

Our investigations of these properties of metals at temperatures 50 - 100°C above the melting point fail to provide any justification for such a conclusion; this can be seen from the data of Table 21, which gives the values we have obtained for surface tension, viscosity and fluidity of the purest metals.

Thus, other pouring conditions being the same, the value of the fluidity of pure metals does not depend on their surface tension. Also, no regular connection can be seen between viscosity and fluidity. The fluidity of metals is determined by the combination of the thermophysical properties of the metals: heat of crystallization, specific heat, and thermal conductivity. The heat of crystallization is the main factor: the greater it is, the higher is the fluidity. Thermal conductivity acquires material significance if crystalline formations or casting crusts are present in the melt, since the thermal conductivity varies abruptly when metals pass from the liquid to the solid state; it is to be assumed that the higher the thermal conductivity of a metal, the lower will be its fluidity. The absence of any direct connection between the properties of a liquid metal and its fluidity led to the necessity for an investigation of the connection between the fluidity and velocity of flow of a liquid metal in the spiral mold, i.e., outside the influence of factors associated with the crystallization of the metal.

TABLE 21. Comparison of Fluidity of Metals with Their Surface Tension and Viscosity

Metal	Fluidity, cm	Surface tension d/cm	Viscosity	
			Centipoises	Centistokes
Zinc	110	750±20	2.4-2.6	0.34-0.36
Aluminum. .	95	860±20	1.1-1.2	0.44-0.46
Tin	80	525±10	1.8-1.9	0.27-0.29
Lead	68	410±5	2.4-2.6	0.23-0.26

A modification form of spiral was used for determining the velocity of flow of metals; in the determination of fluidity, the cross section of the spiral had the form of a trapezoid with an area of 25 mm^2, while for the determination of the velocity of flow, the cross section was semicircular in form and its area was 6.3 mm^2. This spiral terminated in an orifice 4 mm in diameter, through which the metal flowed freely from the mold, after passing through a distance of 118 cm from its entry into the spiral. The time from the commencement of admission to the end of flow was determined by a stopwatch, the hydrostatic pressure, with which the metal entered the mold and flowed along the spiral was the same, since the weights of the metals tested were the same. The mold temperature was equal to the temperature of the liquid metal. Knowing the weight of the metal, its density and the time of flow, the volume of metal flowing in unit time and the mean rate of flow can be determined.

TABLE 22. Characteristic of Velocity of Flow and Fluidity of Metals

Metal	Temperature, °C	Velocity of flow, cm/sec	Fluidity, cm
Zinc	453	10.5	110
Tin	257	15.17	80
Lead	357	14.42	68

The results of these experiments (Table 22) showed that no definite connection can be found between fluidity and velocity of flow; thus, tin and lead, which have a higher velocity of flow than zinc, have a lower fluidity. Comparison of the data of Tables 21 and 22 indicates that there is a direct connection between viscosity and velocity of flow of metals, since for example for the more viscous zinc, a lower velocity of flow is obtained than for the less viscous tin and lead.

We obtained similar results in the determination of the rate of flow of zinc and tin through a vertical capillary 1 mm in diameter and 40 mm in length. For equal superheat and hydrostatic pressure, the rate of flow of liquid zinc was 0.70-0.71 cc/sec and tin 0.87-0.88 cc/sec, corresponding to the lower values of surface tension and viscosity of tin.

If, however, these values are compared with the values for the fluidity of the metals, no direct relationship can be found between these properties. Properties which are characteristic of metals in a clearly liquid state exert no direct influence on fluidity, i.e., on the ability of metals to fill a mold, if we except the filling of capillary passages. The thermophysical properties of the metal and their variation during its solidification are evidently the principal factors involved in the estimation of fluidity.

On passing to a discussion of the fluidity of alloys, it is to be expected that the filling of a mold by the liquid and crystallizing metal will offer a more complex picture, since the difference in character of the crystallization of pure metals and alloys is very considerable, especially in the case of alloys which solidify over a wide temperature range. These points have been discussed in Chapter I. It is here merely necessary to emphasize the fact that the thermal effects in the crystallization of alloys occupying different positions in the constitutional diagrams may be principally produced sometimes in primary crystallization and sometimes in secondary crystallization.

We have studied the fluidity of certain series of alloys of the following binary systems of eutectic type: Al–Si to 18% Si, Al–Mg to 36% Mg, Pb–Sb to 35% Sb, Mg–Al to 50% Al, Al–Zn, Zn–Sn, and Zn–Pb.

All these alloys were investigated under respectively identical thermal conditions. Before being poured into the spiral, the alloys were heated to temperatures exceeding by 5% the liquidus temperature on the absolute scale.

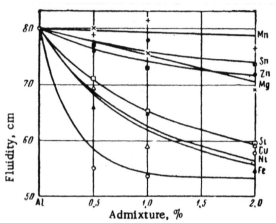

Fig. 65. Influence of small additions of metals on the fluidity of aluminum.

The mold was heated to a temperature 15% below the temperature of crystallization of the eutectic on the absolute scale.

It should be noted that the fluidity values for the pure components of these alloys will be less than those given in Table 17, since in these cases, the temperature of the mold was lower. Anticipating somewhat, we shall point out that this can only emphasize more strongly the high fluidity of the pure metals in comparison with alloys solidifying in a temperature range.

In studying the influence of small additions on fluidity, there is no necessity to vary the thermal conditions of the experiment for the alloy and pure metal, since such additions alter the melting point of the pure metal quite insignificantly (by 2 - 5°C) and produce no appreciable volume of secondary crystallization. A series of experiments was carried out to elucidate the influence of the principal alloying additions on the fluidity of pure aluminum (the experiments were made in the thermal conditions given in Table 17). The data given in Fig. 65 show that in most cases, the fluidity of the aluminum is considerably reduced by the action of small additions of the other component.

The extent of the effect of a given addition on the fluidity of a pure metal is in agreement with the reduction of the heat effect of primary crystallization and the thermal conductivity of alloys. The essential part, however, is played in this case by the character of the structure which is being formed; for instance, in the case of the addition of iron and nickel, the rapid reduction in the fluidity of aluminum is due to the increased ramification of the dendrites of primary aluminum and the occurrence of needlelike crystals of FeAl₃ and so forth. The presence of silicon

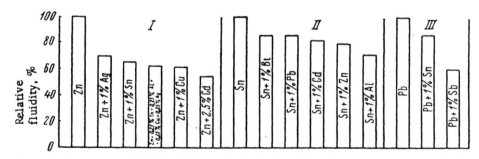

Fig. 66. Influence of small additions on the fluidity of zinc (I), tin (II) and lead (III).

and copper in aluminum produces an appreciable crystallization range (about 100°C), which has a strong influence on the solidification of the alloys and the resulting structure, whereby the fluidity is reduced. Manganese reduces the fluidity of aluminum very slightly, due to the narrow crystallization range and the consequently weak development of dendritic segregation.

The fluidity of zinc, tin, and lead is invariably lowered by the effect of any of the additions tested (Fig. 66). Tin and cadmium have the most pronounced effect in zinc. The presence of tin in zinc results in the crystallization of the alloy over a wide temperature range, and consequently in an increased branching of the dendrites, as is found in actual practice (Fig. 67).

Thus, the reduction of the fluidity of pure metals by the effect of small additions is due to dendritic crystallization and the reduction in the thermal effect of primary crystallization on transition from the pure component to

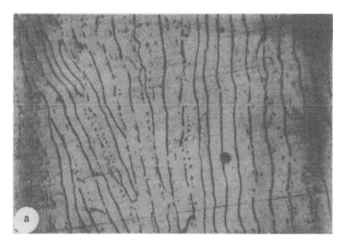

Fig. 67. Microstructure of alloys of zinc and tin (× 150): a) 2.5% Sn; b) 40% Sn.

the alloys. The variation in fluidity with further alloying is caused by both this circumstance and the structure.

The dependence of the variation of fluidity of alloys on their composition, which we have found, is shown in the diagrams of Figs. 68 - 74. The "composition – fluidity" curves in these figures are compared with the form of the constitutional diagrams, since the basic factors determining the variation in fluidity of alloys can be assessed only by means of such comparison.

The general character of the variation in fluidity shows a direct connection between this property and the constitutional diagram, since the minima and maxima on the "composition – fluidity" curves correspond to definite regions or critical points of the constitutional diagrams. Thus, a minimum value of fluidity corresponds to alloys solidifying over the maximum temperature range of crystallization, while the maxima correspond to concentrations at eutectic points and chemical compounds; this latter is to be seen in the diagrams of Figs. 71 and 72.

By the example of alloys of eutectic type, it is possible to examine the influence of factors such as the heat of crystallization which, as was shown above, plays a decisive part in the estimation of the fluidity of pure metals (see Table 17 and the diagram of Fig. 61).

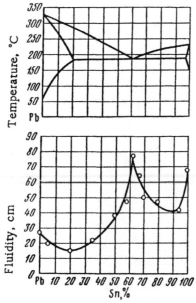

Fig. 68. Dependence of fluidity
of Pb–Sn alloys on composition.

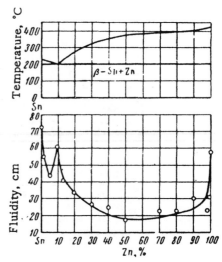

Fig. 69. Dependence of fluidity of
Zn–Sn alloys on composition.

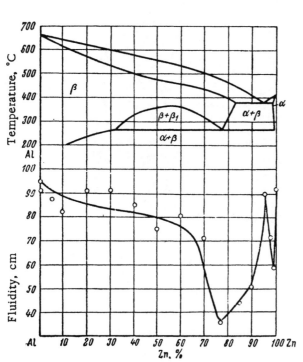

Fig. 70. Dependence of fluidity of Al–Zn alloys on
composition.

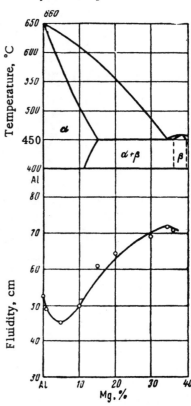

Fig. 71. Dependence of fluidity
of Al–Mg alloys on composition.

In estimating the fluidity of alloys, the relationship of the thermal effects of primary and secondary crystallization must be taken into account. Thus, for example for low-alloy magnesium with 5 - 9% aluminum, the thermal effect of primary crystallization will be less than the thermal effect of the crystallization of pure magnesium. "Temperature-time" curves recorded in the solidification of a number of alloys in the Al–Mg and Al–Si systems (Figs. 75 and 76) show that, for the same cooling conditions, crystallization of the primary phase takes place more slowly than the crystallization of the pure metal or eutectic. In such cases, therefore, the heat of crystallization of the primary formation cannot keep the alloy in the fluid state, while the thermal effect of crystallization of the eutectic will appear only at the instant of final solidification of the casting and will be of no avail.

Thermal conditions are more favorable in the case of magnesium alloys containing 20 - 25% Al. This is due to the fact that the thermal effects of both primary and secondary crystallization assist in keeping the alloy in the fluid state up to find solidification of the bulk of the casting. It is evident that in the crystallization of alloys pertaining to this region of the diagram, the relatively small quantity of primarily separated crystals will move freely in the metal flowing into the mold and will not impede its free flow. The comparatively small thermal effect of the primary separation in this case will be added to the heat of crystallizations of the eutectic, which will produce the favorable conditions for keeping the alloy in the fluid condition until final solidification of the bulk of the casting.

The high fluidity of the alloy of eutectic composition (32% Al) is due to the reserve of superheat heat and the liberated heat of crystallization of the eutectic. This heat, as in the case of the solidification of the pure metal, is continuously liberated during crystallization at constant temperature, up to the solidification of the bulk of the metal.

In view of the fact that the crystallization of eutectics and pure metals proceeds without the growth of the separating crystals, throughout the entire volume of the crystallizing metal, the liquid is able to flow through the channels inside the casting, being "fed" all the time with the heat of solidification of that part of the metal crystallizing immediately on the walls of the mold.

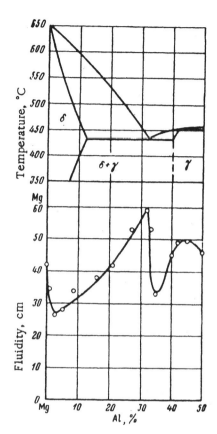

Fig. 72. Dependence of the fluidity of Mg–Al alloys on composition.

Thus, to obtain the same value of fluidity for alloys solidifying in a wide temperature range of crystallization and for the eutectic, a much higher superheat is required in the former case than for the eutectic or pure metal.

This was confirmed directly by experiments with Zn–Sn alloys. It was found that to obtain a fluidity of a zinc alloy with 30% Sn equal to the fluidity of zinc, the alloy had to be heated 50°C higher than the zinc superheated to a corresponding temperature.

Figure 77 shows the microstructure of pure magnesium, indicating that the end of the spiral has a more finely crystalline structure than the commencement of the spiral. Such refinement appears to us to be quite natural, since the metal on reaching the end of the spiral, is enriched in nuclei washed off the already formed casting skin. It is characteristic that in the structure of the specimen taken from the commencement of the cast spiral, crystallization from the mold walls has a sharply pronounced directional character, while at the end of the spiral the structure is equiaxial over the entire cross section, which may be evidence of the simultaneous solidification of this portion of the spiral, with the participation of nuclei washed off the skin.

On the addition of one percent of aluminum to magnesium, the structure of the casting changes abruptly. In the first place, there is no appreciable difference between the grain size at the commencement and end of the casting, and secondly a pronounced grain refinement throughout is observed, compared with the structure of the pure magnesium. At the same time, the fluidity is appreciably reduced. Further increase in the aluminum content of the alloy produces an increase in the crystallization range, which in its turn results in increased dendritization of the structure. The branching dendrites intersect the entire cross section of the casting; it is not difficult to imagine how such dendrites will considerably impede the movement of the liquid metal while it is filling the channel. Such a structure is characteristic of alloys with an aluminum content of 2.5 - 8.5% (Fig. 78).

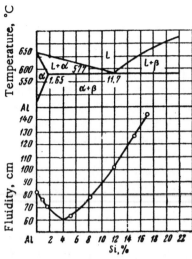

Fig. 73. Dependence of fluidity of Al—Si alloys on composition.

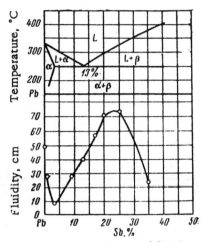

Fig. 74. Dependence of fluidity of Pb—Sb alloys on composition.

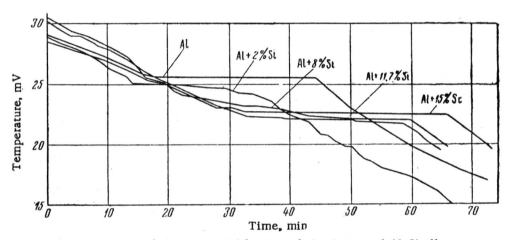

Fig. 75. Time of complete solidification of aluminum and Al-Si alloys.

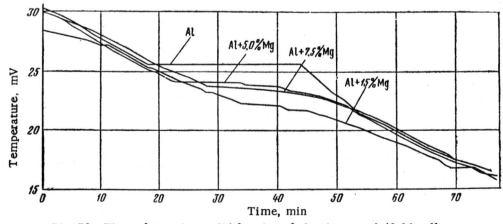

Fig. 76. Time of complete solidification of aluminum and Al—Mg alloys.

The structure of alloys with an aluminum content of 25 - 30% is different. As may be gathered from the structure shown in Fig 79, the primarily formed and growing crystals of magnesium solid solution, during the entire process of flow of the metal in filling the spiral, were isolated from each other, and the eutectic liquid was able to fill the mold more or less easily.

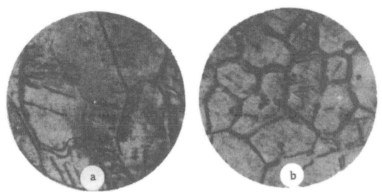

Fig. 77. Microstructure of magnesium (× 90): a) Commencement of spiral; b) end of spiral.

The structure in Fig. 80 shows that the liquid eutectic flows freely and carries within itself only isolated formations of magnesium solid solution. The fluidity of alloys with such a structure is much higher than that of alloys having a lower aluminum content (see the curve in Fig. 72).

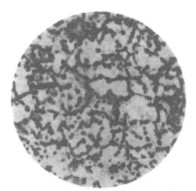

Fig. 78. Microstructure of magnesium alloy with 2.5% aluminum (× 90).

Fig. 79. Microstructure of magnesium alloy with 27% aluminum (× 90).

The structure shown in Fig. 81 is characteristic of alloys of hypereutectic composition; according to the constitutional diagram, an alloy containing 33.5% Al solidifies with the formation of primary crystals of the chemical compound Mg$_4$Al$_3$. In the structure of specimens of the alloy taken from the front end of the spiral, these crystals have the form of long, thin dendrites with incompletely developed branches. The intertwining of such crystals exerts a definite retardation on the flow of the eutectic liquid, and in this case, the fluidity is actually lower than for an alloy of eutectic composition.

Thus, the form of the crystals produced in the initial stage of solidification of alloys has a considerable influence on fluidity; the presence of such crystals will reduce the fluidity to a varying degree (depending on their form), while the heat of formation of the crystals will promote fluidity.

These two factors — thermal and structural — appear very characteristically in Al—Si and Pb—Sb alloys (see Figs. 73 and 74). The fluidity maximum on the "composition — fluidity" curves is shifted from the eutectic point to the region of hypereutectic composition. The observed increase in the fluidity of hypereutectic alloys in such cases is primarily due to the relatively high thermal effects of the primary crystallization of silicon and antimony.

Of the metals (according to certain published data) silicon has the highest heat of crystallization − 327 cal/g, being more than three times the heat of crystallization of aluminum. The heat of crystallization of antimony exceeds still more the heat of crystallization of lead (38.3 against 5.6 - 6.3 cal/g for lead). It is evident that even a small quantity of the primary crystals of these metals, formed during the solidification of hypereutectic alloys, will assist in

Fig. 80. Microstructure of magnesium alloy with 32% aluminum (× 90).

Fig. 81. Microstructure of magnesium alloy with 33.5% aluminum (× 90).

keeping the alloys in the fluid condition. The effect of this factor is combined with a favorable structure, since the separated crystals have an equiaxial form and are isolated from one another. In such a case, they can move about freely in the flowing liquid, only slightly increasing its viscosity.

In Al−Si and Pb−Sb alloys, the compact primary crystals of silicon (Fig. 82) and antimony do not impede the flow of the eutectic liquid, while the primary crystals of the chemical compound Mg_3Al_4, in the form of needlelike undeveloped dendrites (see Fig. 81), produce a retardation of the flow of liquid melt, and the fluidity of the alloys in such a structure is diminished.

From an examination of the connection between the fluidity of Al−Si and Pb−Sb alloys and the form of the constitutional diagram, it may be assumed that the increase in fluidity of the hypereutectic alloys is also connected with the steep rise in the liquidus curve, i.e., the relatively high heating of the alloys above the eutectic temperature, forming their basis. As shown by experiments, however, superheating has no special significance in the present case,

Fig. 82. Microstructure of aluminum alloy with 15% Si (× 90).

while the fundamental conditions determining the high fluidity of these alloys are the heat of crystallization and the form of the primary crystals. Thus, for example, for the same pouring temperature of 620°C, the fluidity of the alloy with 15% Si is 120 cm, while for the eutectic alloy (11.7% Si), it is only 101 cm. This is all the more significant since at such a temperature, the alloy with 15% Si is in the solidification range, i.e., it was poured into the mold at a temperature below the temperature of the commencement of primary crystallization of the silicon. Nevertheless, the fluidity of this alloy was found to be higher than the fluidity of the eutectic alloy, heated above the liquidus temperature. It is evident that in the present case, it is not the superheating heat, but the combined heat of crystallization of the primary silicon and of the eutectic, and the form of the primary crystals which determine the high fluidity of the alloy with 15% Si, cast at a temperature below the liquidus.

In hypoeutectic alloys (for example with 8% Si) in which the aluminum crystallizes in the form of branching dendrites, the value of fluidity for the same pouring temperature is much less (75 cm). When this alloy is poured at a temperature for which the alloy is in the solidification range, its fluidity drops considerably (47 cm).

What has been said above regarding the combined effect of the heat of crystallization and the form of the primarily separated crystals on the fluidity of alloys of the Al—Si system also applies wholly to alloys of the Pb—Sb system.

Fluidity of Alloys of Ternary Systems

The physicochemical basis of the fluidity of alloys, and the direct dependence of fluidity (for identical thermal and velocity conditions of casting) on the form of the constitutional diagram, observed in alloys of binary systems would not be fully demonstrated without experimental confirmation by examples of more complex alloys.

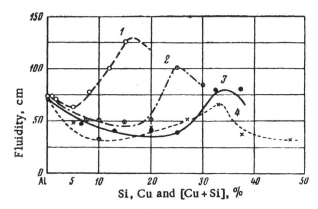

Fig. 83. Dependence of fluidity on composition of ternary alloys of the Al—Cu—Si system: 1) Al—Si alloys; 2) alloys Al—[Cu+Si](Cu : Si = 1 : 1); 3) the same (Cu : Si = 4 : 1); 4) Al—Cu alloys.

To elucidate the phenomena and relationships reflecting the variation in fluidity of complex alloys, "composition − fluidity" diagrams were constructed for alloys of the Al—Cu—Si and Al—Mg—Zn systems, forming the basis of many industrial aluminum alloys; such an investigation for complex alloys had not previously been made. The test conditions were the same as before, i.e., identical relative heating of the liquid metal above the liquidus curve, the same heating of the mold below the solidus temperature.

The results of this work are shown in the diagrams of Figs. 83 and 84 in the form of "composition − fluidity" curves for several radius sections of the diagrams. As was to be expected, the value of the fluidity of the alloys varies regularly with varying quantity of alloying components and with the form of the constitutional diagram, and in particular with the position of the eutectic point. Minima on the fluidity curves correspond to alloys solidifying over a wide temperature range, and maxima to alloys near the eutectic composition. A characteristic feature is also that the value of the fluidity of ternary alloys and its maximum for the eutectic lie between the values of the alloys of the binary systems, which corresponds to a varying quantity of heat of crystallization of the alloys. Thus, for example, the fluidity maximum for alloys on the section Al—(Cu : Si = 4 : 1), on the section near the section aluminum − ternary eutectic (63.5% Al, 31.5% Cu, and 5% Si) is much lower than the maximum for the binary Al—Si alloys, rich in eutectic, and somewhat higher than for the Al—Cu alloys. The reason is that the heat of crystallization of the ternary eutectic (calculated according to the mixing rule) is 93.8 cal/g, while the heats of crystallization of the binary Al–Cu and Al–Si alloys are 78.9 and 129 cal/g, respectively.

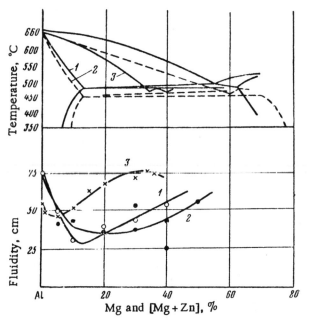

Fig. 84. Dependence of fluidity on composition of ternary alloys of the Al–Mg–Zn system: 1) Alloys Al–[Mg + Zn] (Mg : Zn = 2 : 3); 2) the same (Mg : Zn = 3 : 8); 3) Al–Mg alloys.

Thus, for alloys of ternary systems, the same regular relationship is observed between the fluidity of the alloys and their composition and the form of the constitutional diagram, just as for binary alloys. The principal factors determining the fluidity of ternary alloys are also, on the one hand, the relationship of the heats of primary, secondary, and tertiary crystallization (determined by the physical nature of the components and the quantitative relationship of the phases on solidification in accordance with the position of the alloys in the constitutional diagram, and on the other hand, by the form of the primary precipitated crystals.

Connection Between Fluidity of Alloys and Their Surface Tension

The question concerning the connection between the fluidity of alloys and their surface tension has not formed the subject of any systematic elucidation in the technical literature. One of the reasons is the lack of experimental data on the surface tension of liquid metals and alloys, and its variation on alloying. Available data on the variation of the fluidity of alloys as a function of temperature and certain other factors relate mainly to alloys of a definite composition.

Yu. A. Nekhendzi points out that the surface tension of liquid steel affects its fluidity only when narrow channels of molds are being filled; the lower the surface tension, the more readily does the metal fill these passages. He also points out that surface films and oxides play an important part [80, 150]. A similar conclusion concerning the part played by films, oxides and nonmetallic inclusions is to be found in a number of other publications [7, 165]. In some publications on foundry practice, it is stated that the higher the surface tension, the higher is the fluidity of alloys [157-159]. In [165], on the contrary, one advantage of steel for castings is considered to be its high surface tension, together with high fluidity.

In a paper by V. I. Prosvirin et al., [166] on the atomization of liquid metals, it is stated that the "most regular spherical particles produce alloys having the lowest fluidity, i.e., having high surface tension." O. S. Bobkovaya and A. M. Samarin [129] found an increase in fluidity in the case of chromium-nickel alloys of low surface tension.

In an investigation of the influence of aluminum on the surface tension and fluidity of steel in a U-shaped test-piece, it was found that an addition of aluminum of up to 0.06% increased the values of these two properties simultaneously; further increase in the aluminum content reduced them, especially the fluidity [150]. On the other hand, data are given in the same paper to show that the fluidity of modified silumin, measured in a vacuum test (suction of the metal into a quartz tube) was found to increase with the decrease in the surface tension which occurred when the alloy was allowed to stand after modification.

Thus, earlier work contains contradictory data and fails to provide an adequate concept of the connection between fluidity and surface tension, both of which play an important part in the production of castings from alloys of various compositions.

The results we have obtained on the regular variation of these properties of alloys as a function of composition and in connection with the form of the constitutional diagram enable this problem to be clarified in greater detail. Comparative tests were made of the fluidity of a number of alloys, in which the surface tension on alloying either varies very considerably or does not vary at all. The very first experiments on aluminum and zinc to which surface-active lead, bismuth and antimony were added, showed that the observed considerable reduction in surface tension (see Figs. 29 and 31) does not result in any appreciable change in the fluidity of the alloys.

A systematic comparison of the "composition − surface tension" diagrams with "composition − fluidity" diagrams leads to the following:

1. For a number of alloys, for which the value of the surface tension practically does not depend on composition, the fluidity varies within wide limits, passing through a minimum for alloys solidifying in a temperature range, and through a maximum for alloys of eutectic composition (Al−Si, Al−Cu, Al−Zn and other alloys; see Figs. 29 - 31 and 68 - 74).

2. For a number of alloys, for which the surface tension is reduced appreciably on alloying, fluidity varies according to the curves described above, i.e., having a minimum and a maximum (Al−Mg, Zn−Sn, Pb−Sb alloys; see Figs. 29 - 31 and 69, 71, 74).

3. For a number of alloys, for which the surface tension is considerably reduced on alloying, the fluidity either does not alter of is slightly reduced (Al−Sb, Al−Bi, Zn−Pb and other alloys; see Figs. 29, 31, 33).

4. Of a large number of tests in which the surface tension was considerably increased on alloying, no case was found where there was also found a considerable increase in fluidity.

5. In alloys of the ternary system Al−Cu−Si, for a practically constant surface tension, fluidity varies considerably, depending on the position of the alloy in one region or other of the constitutional diagram. Maximum fluidity is found in alloys near the ternary eutectic point of the diagram (see Fig. 83).

6. Alloys of low surface tension fill the cross section and sharp corners of the spiral better.

Fluidity, as the ability of metals and alloys to fill a mold, is thus associated with a number of physicochemical properties of molten metals and metal solutions. The fluidity of pure metals is mainly determined by the heat of crystallization and thermal conductivity; the greater the heat of crystallization and the lower the thermal conductivity, the higher is the fluidity; the presence of impurities reduces the fluidity of pure metals. The fluidity of binary alloys depends on their position in the constitutional diagram; the "composition − fluidity" curves obtained for compatible temperatures of metal and mold, definitely show that pure components, alloys of eutectic composition and chemical compounds, crystallizing at constant temperature, have maximum fluidity. Alloys solidifying in a crystallization range have relatively low fluidity. The lower fluidity of such alloys, compared with those mentioned above, is due to the relative variation in the quantity of heat liberated in primary and secondary crystallization, to the variation in thermal conductivity of the liquid-solid mixture and to the form of the crystals produced.

Relatively high fluidity is possessed by some binary alloys near the eutectic composition, in which the primary formations have a high heat of crystallization and the crystals of which are small and have a balanced form.

The "composition − fluidity" and "composition − surface tension" diagrams for a number of binary and ternary alloys indicate that the fluidity of these alloys does not depend on the value of their surface tension. A direct connection evidently exists between fluidity and surface tension in those cases where the composition of the alloy corresponds to chemical compounds of constant composition, and to solid solutions, crystallizing at constant temperature (at the minimum or maximum of the melting curves). In such cases, an extreme value of surface tension of the alloys may correspond to maximum fluidity.

The experimental results we have obtained on the regular variation of fluidity of binary alloys with variation in their composition and in connection with the constitutional diagrams are in agreement with the theoretical conjectures made by A. Portevin, A. A. Bochvar, Yu. A. Nekhendzi, A. M. Samarin, B. B. Gulyaev and others. These conclusions refute the conclusion of certain foreign investigators [156, 159] and Soviet investigators [157, 158] to the effect that the fluidity of alloys depends directly on the surface tension. Surface tension is a characteristic of alloys in the definitely liquid state, while fluidity is a characteristic of them in the liquid-solid state.

What has been said about alloys of binary systems applies also to alloys of ternary systems.

It has been shown above that the fluidity of pure metals is not directly connected with such properties of liquid metals as viscosity or velocity of flow. In regard to alloys, there are investigators who find that there is something in common between loss in fluidity of alloys solidifying in a wide temperature range and their viscosity in the liquid state. Thus, the English investigator V. Kondic [156] in a paper on the fluidity of alloys of the Al—Si system explains the reduction in fluidity of low-concentration alloys as being due to the increase in their viscosity in the liquid state, which was also found to be the case by the Soviet workers É. V. Polyak and S. V. Sergeev. In a subsequent study of this problem by É. G. Shvidkovskii, their results were refuted by a detailed check, which showed that the viscosity values of alloys obtained by É. V. Polyak and S. V. Sergeev were too high, due to imperfections in the method of determination [41].

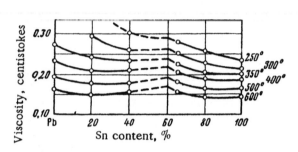

Fig. 85. Kinematic viscosity of Pb—Sn alloys.

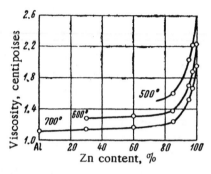

Fig. 86. Viscosity of Al—Zn alloys [164].

On the basis of the results of E. G. Shvidovskii's work on the viscosity of alloys of the Pb—Sn system, we may say that the viscosity of liquid alloys definitely does not determine their fluidity, this being shown very well by a comparison of the diagrams of Figs. 85 and 68, in which the variation of these properties has been plotted as a function of the composition of the alloys. The viscosity of the alloys does not vary appreciably, while the "composition — fluidity" curve has a maximum at the eutectic point. Furthermore, for the high-lead alloys, an inverse relationship is found for these properties, the increase in lead in the alloys resulting in some reduction in viscosity, while the fluidity is considerably reduced. At the same time, the alloy of eutectic composition with a somewhat higher viscosity in comparison with the adjacent alloys shows maximum fluidity.

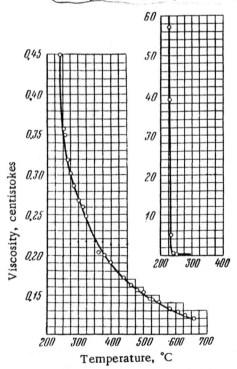

Fig. 87. Kinematic viscosity of the alloy with 75% Pb + 25% Bi. The diagram in the top right-hand corner shows the viscosity of this alloy in the heterogeneous region of the constitutional diagram.

A similar lack of agreement in the character of the variation of these properties is also found in the case of Zn—Al alloys, as may be seen from a comparison of the "composition — fluidity" curve (Fig. 70) and the "composition — viscosity" curve (Fig. 86). In this case, when zinc is alloyed with aluminum, the viscosity of the liquid alloys falls continuously, which ought to increase their fluidity. In fact, however, the fluidity of the alloys in this range of concentration first of all decreases, then increases, reaching a maximum value at the eutectic point and then decreases again, just as was found in all our other investigations of binary alloys of eutectic type.

For alloys of the Al—Si system in the hypereutectic region, a rise in the fluidity curve continuing after the eutectic is characteristic (see Fig. 73), although the viscosity increases at the same time (see Fig. 56). A comparison of these properties for alloys of the Zn—Sn system (see Figs.

69 and 59) also shows convincingly that the fluidity of the alloys is not directly connected with their viscosity: the "composition – fluidity" curve falls sharply with the alloying of these metals and has a maximum at the eutectic point, while the viscosity curves vary steadily.

These contradictions are removed by making use of E. G. Shvidkovskii's results on the viscosity of alloys in the heterogeneous region, i.e., when crystals are present in the liquid. Even an insignificant drop in temperature below the liquidus curve results in a considerable increase in viscosity (by a factor of several tens or hundreds), due to the appearance of primary crystals in the liquid, as may be gathered from the viscosity curve of Pb–Bi alloys for temperatures in the transition range (Fig. 87).

We studied the velocity of flow of a number of liquid alloys of a composition near the eutectic point, where fluidity varies very rapidly. The manner in which these experiments were made was the same as that for the determination of the velocity of flow of pure metals (see p. 74). The temperature of the metal and mold (spiral) was the same, being equal to the temperature of the metal in the determination of its fluidity. The character of the variation of the velocity of flow and fluidity with the composition of Pb–Sn and Pb–Sb alloys is shown graphically in Fig. 88. As will be seen, no variation whatsoever occurs in the velocity of flow (viscosity) of the liquid alloys with variation in composition, while the fluidity varies by a factor of 1.5 - 2. These results show that no definite connection is found between properties of alloys in the liquid state such as velocity of flow (a quantity which generally speaking is the reciprocal of viscosity) and their ability to fill molds. Such a connection may evidently be realized mainly in cases where the viscosity of alloys in the heterogeneous region of the constitutional diagram is considered, i.e., during solidification in a temperature range. Consequently, the existing opinion that when filling ordinary molds, the metal is in a definitely liquid state is not always justified. The observed facts, particularly the high fluidity of hypereutectic Al–Si and Pb–Sb alloys, which are capable of filling a mold (spiral) in the liquid-solid state, is rather evidence in support of the opposite view.

<u>Fluidity of Industrial Alloys in Relation to Casting Conditions and</u>
<u>Position of the Alloys in the Constitutional Diagrams</u>

In the foregoing, the fundamental relationships in the variation of the fluidity of alloys of binary and ternary systems as a function of composition and the form of the constitutional diagram were discussed, and the relationship between these properties of alloys and their thermophysical and other properties was shown. It was found that the fluidity of alloys is practically independent of surface tension.

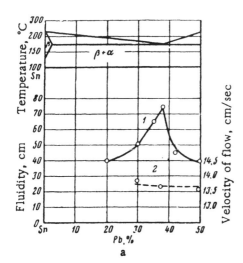 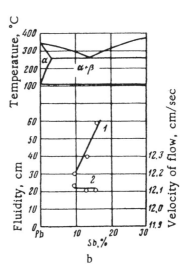

Fig. 88. Variation of fluidity (1) and velocity of flow (2) of alloys near the eutectic composition: a) Sn–Pb alloys; b) Pb–Sb alloys.

With regard to industrial alloys of complex composition, their fluidity may be assessed by comparison with the other alloys on which they are based (for example Al–Si alloys for complex silumins), and for which the "composition – fluidity" curve and the constitutional diagrams are known.

Before proceeding to illustrate this situation, it is necessary to point out the part played by the melting and pouring conditions of the alloys. Fluidity, i.e., the ability of the metal to fill the mold, is affected not only by the temperature-velocity casting conditions, but also very largely by the quality of the liquid metal. It is here primarily necessary to bear in mind that the presence of solid oxides in the liquid metal will considerably reduce the fluidity, in the same way as has been shown for a number of alloys, in which primary crystals of unfavorable form considerably reduce the fluidity.

In fact, as found by Yu. A. Nekhendzi, the fluidity of steel [80], cast iron and aluminum [150] is impaired by repeated remelting. He also gives results [150] showing that a high amount of aluminum introduced as deoxidizer also reduces the fluidity of chromium-nickel-molybdenum steel. Numerous data on the influence of technical factors on the fluidity of steel and the influence of alloying elements (Mn, Si, Cr, V and others) also show that the fluidity of ordinary steel is reduced somewhat on oxidation, this being observed at temperatures of 1600°C and above [82]. It is also pointed out that the addition of aluminum in amounts of 0.2 and 0.6% does not appreciably affect the fluidity, but the latter is considerably reduced on the addition of 1.2%. According to the observations of A. P. Popov, an abrupt decrease in the fluidity of ordinary steel occurs with an addition of aluminum of more than 1.6 - 2.0 kg/ton [144].

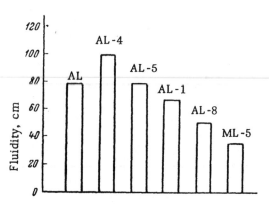

Fig. 89. Fluidity of aluminum and some of its alloys.

Data on the fluidity of industrial alloys on a copper basis [167] are in good agreement with the relationships found for the variation of fluidity in alloys. Thus, with identical thermal casting conditions of the alloys, aluminum bronzes are found to have the maximum fluidity (length of spiral 80 - 100 cm), while in the case of tin-lead bronzes it is very low (20 - 40 cm). The fluidity of brasses of different sorts (L90, L68, L62, etc.) is intermediate between the above-mentioned two groups of alloys (60 - 65 cm). This is due to different methods of solidification of these alloys, as determined by their position in the constitutional diagram. Aluminum bronzes solidify in a very narrow range (15 - 20°C), tin bronzes in a range of 100°C and brasses in a range of 20-50°C.

The fluidity of aluminum-based industrial foundry alloys, the compositions of which mainly lie within the limits of the Al-Si, Al-Cu, Al-Cu-Si and other alloys we have studied, is also completely characterized by the position of these alloys in the constitutional diagrams of the systems mentioned and by the form of the "composition - fluidity" curves. Fig. 89 shows the results of a comparative determination of the fluidity of the commonest standard aluminum foundry alloys. The arrangement of the alloys in order of fluidity is in good agreement with the curves of the variation of fluidity of binary and ternary alloys shown in Figs. 68 - 73, 83, and 84. Thus, the alloy AL-4 in composition near the eutectic, has maximum fluidity, since it crystallizes in a very narrow temperature range (5 - 15°C) and in addition with a favorable relationship of the heats of crystallization of primary and secondary formations, as is shown by the curves of Fig. 61. The aluminum alloy AL-8 with 10% magnesium and the magnesium alloy ML-5 with 9% aluminum, crystallizing in a wide temperature range (100 - 150°C) have the lowest fluidity. It is obvious that these alloys must be strongly superheated to obtain the same value of fluidity as that possessed by the alloys AL-4 or AL-5. We find the same relationship in the fluidity values of alloys of different compositions in the departmental standard (of the aircraft industry) for aluminum and magnesium casting alloys [148]. The agreement between these data and the results obtained in our work would be still more complete if the standard gave the values for the fluidity of the alloys heated not to the same temperatures but to temperatures of equal superheat above the liquidus, as was done in our work.

Also characteristic of industrial aluminum alloys is the drop in fluidity with increase in the quantity of oxides in the melt, for example when turnings are used for melting. The refining of such melts by means of manganese chloride, however, results in a very appreciable increase in fluidity on casting and an improvement in the mechanical properties [168].

The application of the relationships we have obtained to the analysis of the casting properties of alloys — fluidity, hydraulic tightness, etc. — is also confirmed by data relating to foreign practice. Thus of 22 light casting alloys quoted in the British Standard Specification for 1955 [169], the eutectic silumin with 11 - 13% Si has the best fluidity; the presence of copper, iron or other impurities in an amount of up to 1% of each impairs the fluidity of silumin. These facts follow directly from the "composition - fluidity" diagrams in Figs. 65, 73, and 89. A number of alloys of the British Specification containing 6 - 8% Si have a lower fluidity than the eutectic alloy; the replacement of silicon by copper in the limits of 1.5 - 3% reduces the fluidity of these alloys, similar to what was found in our work (see Figs. 73 and 83). The British Specification also points out that the lowest fluidity is found for the alloys Al–Mg (3 - 6 and 9.5 - 11% Mg), Al–Cu (4 - 5 and 9 - 10% Cu) and also for Al–Cu–Si alloys containing 3 - 5% Cu and up to 3% Si; this statement is in complete agreement with our results as given in Figs. 71, 83, and 89.

Thus, on the basis of the regular curves obtained for the dependence of the fluidity of a number of binary and ternary alloys on their composition and the form of the constitutional diagram, definite conclusions can be made regarding the fluidity of many foundry alloys, either employed in practice or proposed, and the conditions of temperature and speed of pouring them in the molds.

In conclusion, it is necessary to emphasize once more the fact that the fluidity of industrial alloys cannot be related to their surface tension, since comparison of a number of "composition – surface tension" and "composition fluidity" diagrams, covering the composition of many industrial alloys, does not provide sufficient justification for this

Chapter VI

SHRINKAGE PHENOMENA IN ALLOYS AND SOLIDIFICATION CRACKING

In the assessment of the quality of castings and the individual stages of their production, in addition to structure, density, etc., an important part is played by the shrinkage cracks formed in different portions of castings during crystallization and subsequent cooling. In some cases, for example in the continuous casting of steel and other alloys, the possibility of using this modern method of producing high-quality ingots is practically determined by crystallization cracking (hot cracking) [12, 13, 170, 171]. The production of castings in sand and metal molds from all kinds of alloys, and the development of special casting methods, such as the use of permanent molds, precision casting, the investment process and so forth, also require that special attention should be paid to the study of shrinkage as one of the principal properties shown by alloys during their solidification and subsequent cooling.

The lack of systematic investigations on the nature and mechanism of shrinkage phenomena, including the relationships expressing the variation of shrinkage of alloys as a function of their composition and the form of the constitutional diagrams increases the difficulty of combating such forms of waste in foundry production as hotshortness, shrinkage porosity, porous zones, etc.

It is most important for the theory of alloys and for foundry practice to determine the part played by shrinkage phenomena in the crystallization of alloys in a temperature range close to the solidus. It must here be remembered that in addition to crystallization cracking, there are other shrinkage defects, which are characteristic of alloys of given composition, determine in many ways the mechanical properties of the final casting and are not reflected in the value of linear shrinkage; in the presence of slight shrinkage, the mechanical properties of the castings may be very low, and vice versa. Thus, "traditional" bronze with 6% Sn has the least shrinkage of all industrial copper alloys, but its tensile strength does not exceed $20 - 25 \text{ kg/mm}^2$. At the same time, the strength of cast aluminum bronze, with 8 - 10% Al and giving only slight shrinkage, attains $40 - 50 \text{ kg/mm}^2$. If these two bronzes are subjected to machining by pressure followed by annealing, the strength of the tin bronze is doubled, while the strength of the aluminum bronze is increased only by 20 - 25%. Without entering into an explanation of the function of the alloying additions, tin and aluminum, to these bronzes, it may be assumed that the doubling of the strength of the tin bronze by machining under pressure is due to the consolidation of the shrinkage defects produced in casting during crystallization.

The shrinkage when an alloy is cast in a sand mold will be less and the mechanical properties will be lower than when the same alloy is cast in a metal mold. The difference in the magnitude of shrinkage in these cases is a manifestation of the crystallization rate. The more rapidly a casting solidifies, the less development there is in it of physical inhomogeneity, affecting the shrinkage and strength of the metal.

When we speak of the shrinkage of metals, we usually divide it into shrinkage in the liquid state, shrinkage during solidification and shrinkage in the solid state. The first form concerns volumetric contraction and results in a lowering of the level of the liquid metal in the mold. The second form — shrinkage during solidification — in the view of some investigators (for example, [80]), is also only volumetric contraction; it is caused by the change in volume of the metal as it passes from the liquid to the solid state; it is assumed that this form of shrinkage occurs between the liquidus and solidus temperatures [172]. The characteristic of the third form of shrinkage is linear contraction, which appears as a reduction in the linear dimensions of the solidified casting in the cooling process. Casting defects such as hot cracks or tears, are usually associated with the contraction of the solidified metal in the mold. The value of linear shrinkage is usually assumed to be equal to one-third that of the volumetric shrinkage.

Due to their close association with practical foundry production, problems of linear shrinkage have long engaged the attention of investigators. Because of the lack of a proper concept of the crystallization mechanism of metals and alloys occupying different positions in the constitutional diagrams, however, it was some time before a correct solution for these problems was forthcoming.

Without dwelling on early work by British investigators (Turner, Murrey, and Haughton, Turner and Chamberlain and others [173]) on the determination of the linear shrinkage of some metals and Cu—Zn alloys (brasses), Cu—Sn and Cu—Al alloys (bronzes) and others, it will be pointed out that in accordance with the views of F. Sauerwald[115], the linear shrinkage of alloys is due to their thermal contraction with fall in temperature. If a casting cools uniformly in the mold, this contraction should be equal to the sum of the changes in length on the cooling of the metal from the melting point to the temperature of the surroundings, calculated according to the coefficients of contraction in the solid state. Inequality between shrinkage calculated in this way and its actual value, according to Sauerwald, is due to uneven cooling of the casting and also to hydrostatic pressure, set up in the liquid metal during solidification and opposing the shrinkage. It would seem, however, that such a pressure can only occur in the production of large castings.

F. Sauerwald makes no distinction between the solidification processes of pure metals on the one hand and alloys on the other, which is of essential significance in estimating the shrinkage of alloys, solidifying in a temperature range. It is in the case of just such alloys that there is a wide discrepancy between the calculated and experimental shrinkage values, owing to the high susceptibility of alloys to the resistance offered by the mold.

G. Sachs [174] gives data, according to which the shrinkage of pure metals ought to be equal to the thermal contraction of the metal in the temperature range between the melting point and room temperature. In the same

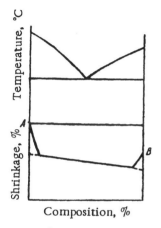

Fig. 90. Theoretical dependence of linear shrinkage of alloys on their composition.

publication, however, Sachs ovserves that this is contradicted by the difference in the shrinkage coefficients of the same metal when cast in a metal mold and in sand. He explains the low shrinkage values, characteristic of alloys, as being due to the existence of preshrinkage expansion on solidification. This view of G. Sachs does not seem to be entirely justified, since not every alloy is subject to preshrinkage [175].

Yu. A. Nekhendzi [80] and P. I. Bidul' [172] associate volumetric (liquid) shrinkage and linear (solid) shrinkage with the thermal expansion of the metal in the liquid and solid states, respectively, and also with the superheat temperature and solidus temperature of alloys.

The investigations of Academician A. A. Bochvar on the theory of the linear shrinkage of alloys have shown that the thermal contraction of metals and alloys, as determined numerically from their coefficients of expansion, is the only factor affecting the magnitude of linear shrinkage. For pure metals and alloys crystallizing at constant temperature, linear shrinkage appears only after solidification, and the value of the shrinkage must be equal to the product of the coefficient of expansion by the value of the melting point. For alloys crystallizing in a temperature range, linear shrinkage sets in after the formation of a continuous crystalline skeleton in the solidifying volume, and this may also take place in the presence of a certain quantity of residual liquid in it [176]. The curves proposed by A. A. Bochvar for the linear shrinkage of alloys are reproduced in Fig. 90. Such a relationship of linear shrinkage was confirmed in a number of actual alloys,

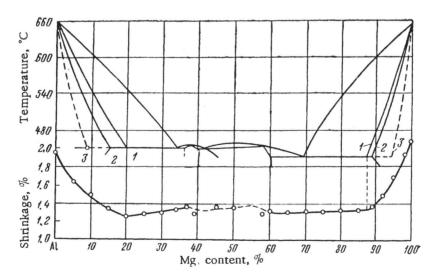

Fig. 91. Dependence of linear shrinkage of Al–Mg alloys on their composition [177]: 1) Temperature of commencement of linear shrinkage; 2) solidus in equilibrium conditions; 3) "nonequilibrium solidus".

and it was found that the linear shrinkage of alloys solidifying in a temperature range commences from a certain temperature, the position of which is indicated in the constitutional diagram of Fig. 91. The portion of the diagram between this line and the solidus line was called the effective crystallization range.

In subsequent investigations, A. A. Bochvar points to the importance of this range in assessing such properties of alloys as resistance to shrinkage stresses [40, 176]. The theoretical curves of linear shrinkage of eutectic alloys in Fig. 90 are based on the concept that a portion of the shrinkage of alloys, which are close to the pure components in composition, cannot be realized, due to loss of contraction of the primarily separated formations on crystallization in the temperature range from the liquidus to the curve of the commencement of linear shrinkage. After the curve of the commencement of linear shrinkage has intersected the eutectic horizontal, the magnitude of the shrinkage of the eutectic mixture will vary according to the additivity rule, i.e., according to a straight line. The slope of this straight line will depend solely on the difference in value of the coefficients of thermal contraction of the phases forming the eutectic mixture, since the commencement of the line of linear shrinkage of alloys remains constant, being situated on the eutectic horizontal.

It follows from an examination of the curves in the variation of linear shrinkage of alloys with their composition that of a series of eutectic alloys, minimum shrinkage will be found in alloys near the component with a low coefficient of expansion. This work refutes the earlier view held by Wüst and others to the effect that alloys corresponding to the eutectic composition have minimum shrinkage [176].

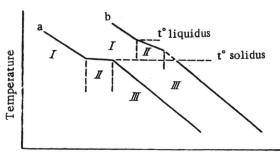

Fig. 92. Variation of volume of metals and alloys with temperature. a) Shrinkage variation of volume of pure metals, eutectics and chemical compounds, solidifying at constant temperature; b) the same for alloys solidifying in a wide temperature range; I) in the liquid state; II) during solidification; III) in the solid state.

We shall examine in what follows the relationship between the linear shrinkage of metals and their thermal characteristics, and also the influence of composition on shrinkage phenomena. The solidification conditions of alloys will be considered in relation to the position of the alloys in the constitutional diagrams, which in our view is the most correct physicochemical basis for establishing the relationships in the variation of the different properties of alloys, including such properties as shrinkage and susceptibility to cracking during crystallization in hindered shrinkage conditions.

A detailed study of the mechanism involved in the transition of alloys from the liquid to the solid state is essential for ascertaining the actual causes of cracking in castings and its prevention. Knowing the character of the crystallization of alloys occupying different positions in the constitutional diagrams, we can assess correctly the part played by shrinkage phenomena and the influence of linear shrinkage on the production of solidification cracking, independently of the action of external factors.

It should be stated that the above-mentioned definition of volumetric shrinkage as being due solely to the fall in level of the liquid metal will apply to crystallization of pure metals and alloys close to the eutectic. Obviously, for alloys crystallizing at variable temperatures in a wide range, solidification shrinkage can be produced only partly in the form of a fall in level of the liquid metal, since the free access of the latter to points where solidification is proceeding does not occur equally in all alloys, being particularly difficult in those alloys where branching dendrites permeate the mass of the alloy. In such cases, linear shrinkage commences already while solidification shrinkage is still in progress, although crystallization of the entire mass of the alloy has not yet been completed.

The general form of the variations in dimensions of a metal cast in a mold are shown in Fig. 92. It is evident that for pure metals, eutectics and other alloys solidifying at constant temperature, the change in volume at the point of transition will be characterized by a horizontal portion (Fig. 92,a) and for alloys crystallizing in a temperature range (Fig. 92,b), passage through this range will be accompanied by a drop in level of the liquid, due to its drop in temperature and the transition of part of the liquid to the solid state (solidification shrinkage), and also due to the free contraction of the separating crystals completely surrounded by liquid. For certain alloys with maximum solidification range, the uniform fall in level of the liquid metal in the casting may cease and the linear shrinkage of the growing dendrites may commence. In our view, this forms the principal difference between character of the shrinkage of pure metals and that of some alloys solidifying in a wide temperature range.

Table 23 gives data on the increase in density of certain metals on solidification and the decrease in density on heating in the liquid state.

Agreement between the fundamental theoretical concept of the direct dependence of the shrinkage of metals on the coefficient of expansion and melting point has been verified for a number of metals. The method of determining the shrinkage has been described previously [180]. The experimental results, as a mean of three or more determinations, are shown in Table 24.

The results given in Table 24 show that the values for the shrinkage of pure metals, obtained experimentally and by calculation, are in close agreement; the differences between them are within the limits of error in the determination of both shrinkage and the coefficient of expansion. Consequently, the coefficient of thermal contraction and the melting point constitute a measure of the linear shrinkage of pure metals. The shrinkage values can be obtained from the equation:

$$\varepsilon = \alpha_t (t_m^\circ - t_{20}^\circ) \cdot 100\%$$

where α_t is the mean coefficient of expansion of the metal in the range from melting point to room temperature.

TABLE 23. Increase in Density of Metals on Solidification and Decrease in Density on Heating in the Liquid State [135, 179]

Metal	Density at melting point, g/cc		Increase in density on solidification, %	Decrease in density on heating in the liquid state by 1°C
	Solid phase	Liquid phase		
Ag	9.7	9.3	4.3	0.0009 (960—1300°)
Al	2.56	2.4	6.6	0.00028 (660—110°)
Bi	9.7	10.0	—3.0	0.0011 (300—962°)
Cd	8.4	8.0	5.0	0.00108 (330—600°)
Cu	8.35	8.0	4.4	— —
Mg	1.65	1.57	5.1	0.00108 (651—750°)
Pb	11.0	10.7	2.8	0.00117 (400—1000°)
Sn	7.2	7.0	2.86	0.00064 (409—704°)
Zn	6.9	6.6	4.54	0.00092 (419—800°)

TABLE 24. Shrinkage of Metals Determined Experimentally and by Calculation (from the Value of the Coefficient of Expansion)

Metal	Shrinkage, %		Mean value of the coefficient of expansion in the limits of the temperatures indicated		Shrinkage according to the data of other authors, (experimental), %
	Experimental	Calculated	$\times 10^{-6}$	$t_1 - t_2$, °C	
Ag	1.97	1.96	20.5	0—900	1.97 [181]
Al техн.	1.62—1.73	1.5—1.77	22.65—26.8	0—600	1.78 [182]
	1.62—1.73	1.8	28.2	0—600	1.8 [177]
Al 99,99%	1.96	1.84	28.7	0—600	1.7 [170]
Bi	0.41	0.395	14.6	0—270	0.22 [183]
Cd	1.10	1.22	38.0	0—315	
Cu	2.16	2.12	20.0	0—1000	1.99 [181]
Cu	2.16	2.16	20.3	0—1000	2.1 [167]
Mg	1.87	1.85	29.3	0—600	2.0 [191]
Mg	1.87	1.86	29.6	0—500	0.9 [183]
Pb	1.11	1.08	33.0	0—320	1.1 [185]
Sn	0.63—0.67	0.60	28.5	20—232	0.55—0.44 [186] 0.8 [185]
Zn	1.58	1.36	34.0	0—400	1.60 [185]
Zn	1.58	1.42	35.4	0—300	1.61 [182]
Zn	1.58	1.58	39.7	20—300	1.57 [167]

TABLE 25. Shrinkage of Metals Obtained by Calculation (From the Density or Specific Volume)

Metal	At the boiling point		At 18°C		Calculated shrinkage, %		Experimental shrinkage, %
	Density, g/cc	Specific volume, cc/g	Density, g/cc	Specific volume, cc/g	Volumetric	Linear	
Ag	9.7	0.0103	10.5	0.00952	7.6	2.53	1.97
Al	2.56	0.371	2.69	0.391	5.13	1.71	1.73—1.96
Bi	9.7	0.0103	9.8	0.0102	1.07	0.36	0.41
Cd	8.4	0 119	8.64	0.115	3.36	1.12	1.1
Cu	8.35	0.112	8.93	0.12	6.66	2.22	2.16
Mg	1.65	0.574	1.74	0.606	4.95	1.65	1.87
Pb	11.0	0.0909	11.34	0.0882	3.0	1.0	1.11
Sn	7.2	0.1388	7.28	0.1373	1.22	0.41	0.63
Zn	6.9	0.145	7.14	0.14	3.45	1.45	1.58

TABLE 26. Linear Shrinkage of Zinc and Cadmium

Metal	Coefficient of expansion		Calculated shrinkage, %			Experimental values of shrinkage of polycrystalline specimen, %
	parallel to axis	perpendicular to axis	parallel to axis	perpendicular to axis	Mean	
Zn	63.9	14.1 [179]	2.56	0.564	1.56	1.58
Zn	59.0	16.0 [183]	2.36	0.64	1.50	—
Cd	52.6	21.4 [179]	1.58	0.64	1.11	1.10

TABLE 27. Decrease in the Linear Shrinkage of Metals in the Case of Hindered Shrinkage

Metal	Load during solidification, kg/mm²	Shrinkage, %
Aluminum	Without load	1.85-1.95
	0.012	1.64
	0.025	1.53
	0.042	1.40 (cracks)
	0.084	1.13 (cracks)
Zinc	Without load	1.56-1.58
	0.006	1.30
	0.009	1.20
	0.018	1.19
	0.076	1.10
	0.098	0.09
Tin	Without load	0.64
	0.014	0.61
	0.028	0.57
	0.056	0.52
Lead	Without load	1.11
	0.015-0.020	0.55-0.6

The direct connection between the linear shrinkage of metals and their physical constants is also shown by the data of Table 25, in which the value of shrinkage has been calculated from the density (specific volume) of metals at the melting point and at 18°C. A certain discrepancy between the calculated and experimental values in this case is due evidently to the considerable error in the determination of the density of metals near the melting point. With the exception of silver, zinc and tin, however, the discrepancy does not exceed 15% of the measured value; if the calculation is made according to the coefficients of expansion, this discrepancy does not exceed 6%.

It should be noted that maximum discrepancy between the experimental and calculated values of shrinkage is found for metals with a hexagonal lattice. If, however, the shrinkage of polycrystalline zinc and cadmium is calculated on the basis of data on the coefficients of expansion along the principal crystallographic axes of a single crystal, the mean calculated shrinkage value practically agrees with the experimental value, as will be seen from Table 26.

The results obtained indicate the possibility of using the shrinkage method for estimating the unknown mean value of the coefficient of expansion of alloys, if shrinkage is determined without hindrance (in the conditions described in the author's paper [180]) and the alloys undergo no transformation in the solid state.

The linear shrinkage of metals is decreased by the effect of the resistance of the mold (Table 27). In normal casting conditions, therefore, when the mold offers resistance, a decrease in shrinkage compared with the theoretical value is to be expected. At the same time, it is obvious that a load may be selected such that the entire linear shrinkage will occur in the form of plastic deformations in the solidified metal. In such a case, the value of the specific load will vary between the elastic limit and the tensile strength of the metal.

As shown by V. I. Dobatkin [56], with the very high rate of cooling of ingots of aluminum alloys in continuous casting, the actual diametrical shrinkage of the ingot may exceed the calculated value, due to the elastic and plastic compression of the external layers of the ingot by the action of the internal layers.

Dependence of Shrinkage of Alloys on Their Composition and the Form of the Constitutional Diagram

From the results on the shrinkage of pure metals, it may be stated that impurities in or small additions to any metal of components which do not appreciably affect its melting point and coefficient of expansion ought not to alter the value of the linear shrinkage of the metal. This is confirmed by experiments on the determination of the shrinkage of aluminum and copper with additions of metals which only slightly lower the melting point of the alloy (Table 28).

TABLE 28. Shrinkage of Some Aluminum and Copper Alloys

Alloys	Shrinkage, %	Solidus temperature, °C
Al, high purity	1.96	660
Al, commercial	1.72-1.73	655-645
Al + Fe (up to 2%)	1.85-1.83	654
Al + Cr (up to 2%)	1.86-1.83	654
Al + Ni (up to 2%)	1.8-1.7	640
Cu pure	2.16	1083
Cu + Mn (up to 10%)	2.16-2.05	1083-1000

For alloys crystallizing at constant temperature, linear shrinkage begins to make its appearance after the solidus temperature, commencing for example from the eutectic temperature. For alloys crystallizing in a temperature range, shrinkage is observed in the process of formation of the skeleton of primarily separated crystals in the presence of residual liquid phase. According to the investigations of A. A. Bochvar and V. I. Dobatkin, in Al–Mg, Mg–Al and Pb–Sn alloys, linear shrinkage occurs in the solidification range, when 20 - 45% of eutectic liquid is contained in the crystallizing volume [177].

The diagram of Fig. 93 shows the variation in shrinkage of binary alloys of copper with tin, magnesium, phosphorus and antimony in the concentration limits of 10 - 15% of the added metal. The reduction in shrinkage of the

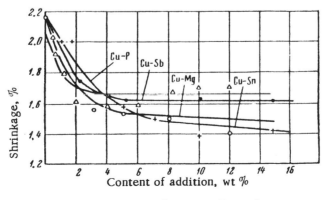

Fig. 93. Shrinkage curves of copper alloys of eutectic type.

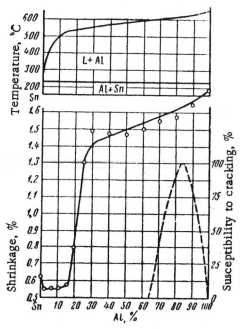

Fig. 94. Dependence of linear shrinkage of Sn–Al alloys on composition (the dashline curve marks the region of alloys susceptible to cracking).

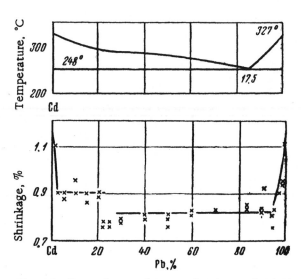

Fig. 95. Dependence of linear shrinkage of Cd–Pb alloys on composition.

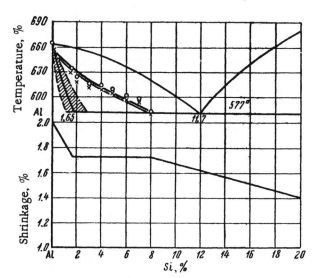

Fig. 96. Dependence of linear shrinkage of Al–Si alloys on composition.

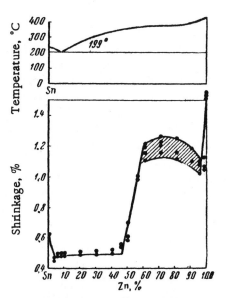

Fig. 97. Dependence of linear shrinkage of Sn–Zn alloys on composition.

alloys corresponds to the drop in solidus temperature of these alloys (compared with the melting point of copper), which for Cu–Sn alloys is 280°C and for Cu–Mg, Cu–P, and Cu–Sb alloys is in the range 360 - 430°C. Maximum reduction in shrinkage corresponds to the limit of solubility of these metals in copper, to which also corresponds the maximum solidification range. In Cu–Sn and Cu–Sb alloys, shrinkage reaches its minimum value before the solubility limit in the constitutional diagram; this is quite conceivable, since the rapid solidification of the small specimens in our experiments led to the occurrence of residual liquid melt at a lower concentration of the second component than that indicated by the constitutional diagram.

The curves reproduced in Figs. 94-100 clearly show the systematic variation in shrinkage of alloys as a function of their position in the constitutional diagram.

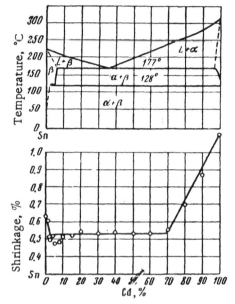

Fig. 98. Dependence of linear shrinkage of Sn–Cd alloys on composition.

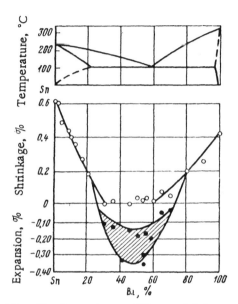

Fig. 99. Dependence of linear shrinkage of Sn–Bi alloys on composition.

In the analysis of "composition – shrinkage" curves, the most interesting portions of the curves are those which slope down from the pure components, due to the sharp drop in the solidus temperature and the formation of a solidification range. The proportion of liquid and solid phases during solidification in a temperature range varies; at a given moment, the mixture of crystals and solid metal begins to show the properties of a solid body, for example a certain strength, linear shrinkage and so forth. The crystallization range may, therefore, be divided into two two-phase regions, in which either liquid or solid crystals predominate.

The fact that the concept of the subdivision of the solidification range into two regions is a correct one has been confirmed by a number of investigations [86, 152, 187, 190]. The physical significance of such a subdivision is made clear by the following experiments. If a paddle stirrer driven by a Warren motor (at a speed of 2 rpm) is placed in a cooling liquid alloy, the motor stops at a certain temperature, since a definite quantity of crystals preventing the rotation of the paddles has been formed. When the solid or solid-liquid mixture is heated again, the motor again begins to rotate the paddles at the same temperature at which the paddles stopped during cooling of the alloy.

The results of such experiments are shown for Al–Si and Sn–Pb alloys (see Figs. 96 and 100), where the experimental points in the constitutional diagram show the starting and stopping of the Warren motor. It will be seen that the motor stops at those alloy temperatures (in slow cooling conditions) for which the alloy contains about 50% solid phase and 50% liquid phase. If with such a proportion of the phases, the alloy has a certain strength opposing the power of the Warren motor, it is obvious that in these conditions there may also be shrinkage of the solid skeleton of the casting, especially in cases where the primary crystals are separated in the form of well-developed dendrites. There is thus an "effective" crystallization range (shaded in the diagram of Fig. 96). Since in this range the solid phase predominates and linear shrinkage occurs, any resistance by the mold may cause the crystals to be moved apart of detached from one another, since they are separated by layers of liquid.

The influence of the quantitative proportions of the phases on the amount of shrinkage during solidification is most apparent in Sn—Al alloys (see Fig. 94). In this case, the aluminum skeleton primarily formed shows its own shrinkage in the liquid-solid region over the entire composition range from pure aluminum to 70 - 80% tin in the alloy (corresponding approximately to 30 - 40 vol. %). For such a tin content, the shrinkage occurring is due to the aluminum, while at higher contents, it is due to the tin.

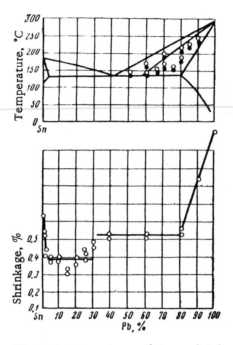

Fig. 100. Dependence of linear shrinkage of Sn—Pb alloys on composition.

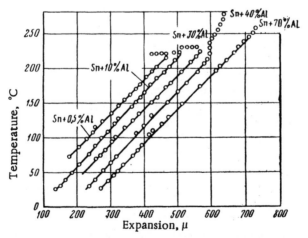

Fig. 101. Expansion curves of Sn—Al alloys on heating to
the eutectic temperature (229°C) and above.

Confirmation that these views on the properties of solid-liquid mixtures are correct may also be found in experiments on the determination of the thermal expansion of alloys of different compositions. The form of the expansion curve of alloys above the solidus temperature enables us to judge whether a solid skeleton, capable of expanding and overcoming the compressive force of the dilatometer spring is present or absent. The results of these observations on Sn—Al alloys are shown in the curves of Fig. 101. If there is no continuous framework of primary crystals in the structure of the alloy, it is capable of expanding only as far as the temperature of the eutectic, after which the specimen begins to contract, as indicated by a sharp turn of the expansion curve in the opposite direction. Alloys in whose structure the primary crystallization products predominate do not contract at the eutectic temperatures under the action of the pressure of the dilatometer spring and continue to expand on further heating at the solidification range temperature.

I. I. Novikov, in an investigation of the hardness and bending strength of alloys at temperatures near the solidus also obtained evidence showing that alloys in the region of the solid-liquid state possess appreciable strength and plasticity [190].

TABLE 29. Linear Shrinkage of Eutectic and Other Alloys

Alloys	Concentration range of alloys with the same shrinkage, %	Eutectic temperature, °C	Shrinkage, %	
			Experimental	Calculated
Al—Si	2—8 Si	577	1.70—1.72	1.5—1.61
Al—Mg	20—35 Mg	451	1.25—1.3 [177]	1.15—1.24
Al—Sn	98—85 Sn	229	0.56—0.58	0.59
Al—Fe	0—3 Fe	658	1.65—1.75	1.70
Cu—Al	2—8 Al	(solid soln.)	2.4	2.0
Cu—P	2—12 P	714	1.6—1.64	1.32
Cu—Mg	3—8 Mg	722	1.53	1.33
Cu—Sn	7—15 Sn	(peritectic, 798)	1.45—1.5	1.54
Cu—Ni	2—15 Ni	(solid soln. 1100—1150)	2.3—2.5	2.15—2.2
Mg—Al	15—35 Al	436	1.3—1.34 [191]	1.23
Mg—Al	7—9 Al	436	1.2—1.4 [142]	1.2—1.25
Mg—Cu	12—20 Cu	485	1.4—1.42 [191]	1.4
Mg—Ni	6—20 Ni	508	1.45 [191]	1.44
Mg—Cd	0—10 Cd	(solid soln. 645—640)	2.0 [191]	1.86
Zn—Sn	95—40 Sn	199	0.5	0.48
Zn—Cd	95—50 Cd	245	0.82—0.85	0.93
Pb—Sn	60—20 Sn	181	0.51—0.52	0.54
Cd—Sn	97—20 Sn	177	0.5—0.52	0.45
Pb—Cd	95—30 Pb	148	0.81	0.75

Thus, when alloys of different compositions are cooled and heated in the solidification range, they exhibit sometimes the properties of a definitely liquid body and sometimes those of a solid. It is obvious that for practical purposes, i.e., for the elimination of defects in castings produced as the result of shrinkage hindrance in solidification, it is advantageous to use alloys in which there is no effective solidification range.

Reverting to the analysis of the "composition – shrinkage" curves shown in Figs. 94-100, attention should be paid to the portions of the curves with constant shrinkage value. It would seem that there ought to be here a change in properties according to the additivity rule, as follows from theoretical considerations [176]. Numerous measurements, however, have definitely shown that shrinkage is invariable over a wide concentration range. It is evident that the shrinkage of two-phase alloys is determined by the component predominating in the given mixture.

If we calculate the shrinkage of alloys for which it is constant (as was done for pure metals, see Tables 24 and 25), it is found that this constant shrinkage value is determined by the eutectic temperature and the coefficient of expansion of the metal quantitatively predominant in the given alloy.

Thus, the following results are obtained for the constant shrinkage values of alloys of the Pb–Cd system: for alloys on a lead basis, the calculated shrinkage will be $33 \cdot 10^{-6}(248° - 20°)100 = 0.75\%$, and for alloys on a cadmium basis it will be $38 \cdot 10^{-6}(248° - 20°)100 = 0.86\%$. Experiment gives 0.80 - 0.82 and 0.88 - 0.90%, respectively (see Fig. 95). The results of these calculations are compared with the experimental data in Table 29.

As will be appreciated from the data of this table, the agreement between experimental and calculated results is good or satisfactory for a number of alloys, including some industrial aluminum-based and magnesium-based alloys. For the solid solution type of alloys (Cu–Al, Mg–Cd) there is no such agreement, since evidently their coefficients of expansion vary with composition more considerably than for eutectic alloys. This also applies to some other alloys given in the table, for example Cu–P and Cu–Mg alloys.

The shrinkage of Al–Fe, Al–Mn and Al–Cr binary alloys with 0.5 – 3% addition does not differ from the shrinkage of commercial aluminum, since the lowering of the temperature of these alloys below the melting point of aluminum amounts to only a few degrees (see Table 28).

The shrinkage data given above for a number of metals and alloys show that alloying with metals producing a depression of the solidus temperature of the alloys most often results in a decrease in shrinkage. This is also true of iron, the shrinkage of which is considerably reduced when it is alloyed with silicon, phosphorus and other elements,

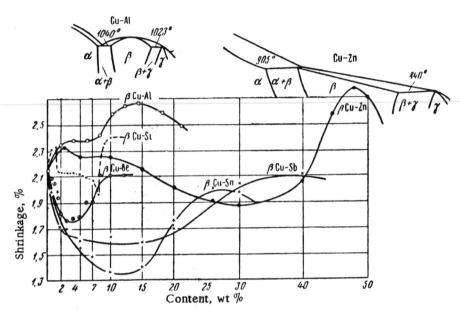

Fig. 102. Dependence of shrinkage of Cu–Al, Cu–Zn and other alloys
on composition and the constitutional diagram (in the β-phase region).

which considerably depress the solidus temperature [80]. For a number of copper-based alloys having a β-structure, however, an increase in shrinkage is found, despite the depression of the solidus temperature (Cu–Zn, Cu–Si and other alloys). This is due to the increase in the coefficient of expansion of the solid solutions, which fully compensates the decrease in shrinkage caused by the existence of a solidification range and the depression of the solidus, and ultimately results in an increase in shrinkage.

Such an influence of the two factors on shrinkage is well illustrated by the "composition – shrinkage" curve of alloys of the Cu–Zn system (Fig. 102). Despite the fact that the coefficient of expansion of these alloys in the α-solution region (10 - 30% Zn) increases with alloying, as shown in Fig. 103, their shrinkage does not increase but even decreases somewhat, since the shrinkage decrease, due to the reduction in the solidus temperature, prevails over the increase, due to the increase in the coefficient of expansion. On passing to alloys with an α + β structure, the coefficient of expansion of which is higher than that of alloys with an α-solution structure, shrinkage is increased considerably (despite a certain depression of the solidus) and reaches a maximum value for alloys having a β-structure.

The character of the variation in shrinkage of alloys in this range of composition is shown in Fig. 102 for a number of other copper alloys. It is here clearly to be seen that maximum shrinkage coincides with the β-phase regions in the equilibrium diagrams of the Cu–Al and Cu–Zn systems. This is also true of the alloys of other systems, the shrinkage curves of which have been plotted in this figure.

It should be stated that the data obtained are in full agreement with the results of the determination of the shrinkage of copper alloys of industrial type, the shrinkage of which is higher than the shrinkage of pure copper, being 2.2 - 2.5% [167].

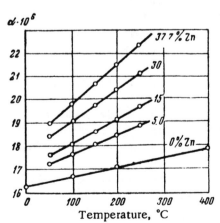

Fig. 103. Thermal expansion curves of Cu–Zn alloys [186].

By making use of the values obtained for the shrinkage of alloys and knowing the solidus temperature, it is possible to calculate the mean coefficient of expansion of the alloys. For alloys of β-phase composition, it was found to be $25 - 27 \cdot 10^{-6}$ (Cu-Al, Cu-Sn and Cu-Be alloys) and $30 - 31 \cdot 10^{-6}$ (Cu-Zn, Cu-Sb and Cu-Si alloys), which is 25 - 50% higher than for pure copper.

It cannot be said that the cause of such a high shrinkage of these alloys is the transformation to the solid state, particularly the eutectoid decomposition of the β-phase, since according to available data this decomposition in copper alloys containing 13.5% Al and 27% Sn occurs with an increase in specific volume up to 0.58 and 0.30%, respectively [193]. In the method we have adopted for the determination of the shrinkage of alloys (casting in a semimetal mold with watercooled bottom), partial decomposition of the β-phase may be expected, which can but result in a reduction in shrinkage. In Cu-Zn alloys with a β-structure according to the same data, a relatively slight decrease in volume (of about 0.16%) is observed at the $\beta \rightarrow \beta'$ transition, and this cannot have any material effect on the linear shrinkage, which for these alloys is 2.6 - 2.75%.

A similar variation in the value of the linear shrinkage of alloys in the $\alpha + \beta$ and β-phase regions is also found in the case of Cu-Zn-Si and Cu-Al-Si ternary alloys, as will be seen from the "composition – shrinkage" curves in Fig. 104; the maxima on these curves are shifted systematically in accordance with the position of the β-phase of these ternary systems, when the composition of the alloys is varied.

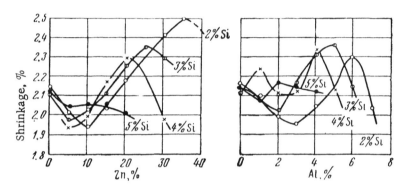

Fig. 104. "Composition – shrinkage" curves of some ternary alloys on a copper basis.

The following are the results of oscillograph recording of the solidification and shrinkage of some alloys. The purpose of these tests was to record simultaneously the temperature and the magnitude of shrinkage at individual stages in the solidification of the alloys. This was done by means of a thermocouple pickup, the junction of which was situated in the solidifying specimen. By switching off the oscillograph lamp for a certain instant at definite moments in the movement of the indicator pointer showing the shrinkage of the alloy, the temperature, time of solidification and the time during which shrinkage assumed a definite value were recorded on the oscillograms. The speed of movement of the oscillograph film was 2 - 3 cm/sec when the specimen was cast in a metal mold and 0.2 cm/sec when it was cast in an asbestos-lined mold. The lamp was switched off every ten divisions of the shrinkage indicator scale, i.e., every 0.1 mm of decrease in the specimen length of 52 mm, corresponding to a shrinkage of 0.192% of the alloy.

Figures 105 and 106 show the more characteristic oscillograms; reading them from right to left, it will be seen that complete heating of the thermocouple to 600 - 700°C by the cast liquid aluminum occurs in 0.5 sec, after which cooling sets in. The beginning and end of crystallization of the metal is marked by a sharp transition of the cooling curve, at the commencement of the process, to a horizontal part and then to a descending part of the curve.

The breaks in the curve at the moment of commencement of shrinkage show that in the solidification of pure zinc and aluminum, shrinkage develops 2.5 - 1.5 sec after pouring and 0.5 sec after the commencement of solidification, when crystallization of the entire volume of the specimen is not yet complete; this is shown by the horizontal part of the cooling curve, continuing after the point indicating the commencement of shrinkage. The recorded shrinkage is the result of the contraction of the crust or skin of the incompletely solidified specimen. Calculation of the curve shows that an increase in shrinkage of aluminum by 0.192% corresponds to a drop in temperature of 55 - 65°C, which is very close to the previously obtained experimental and calculated results (see Tables 24 and 25).

The curves for Zn-Al, Al-Si and Al-Cu alloys show that in the case of alloys, shrinkage commences much later than in the case of pure aluminum or zinc — only 3 - 4 sec after pouring (2 - 2.5 sec after the beginning of solidification).

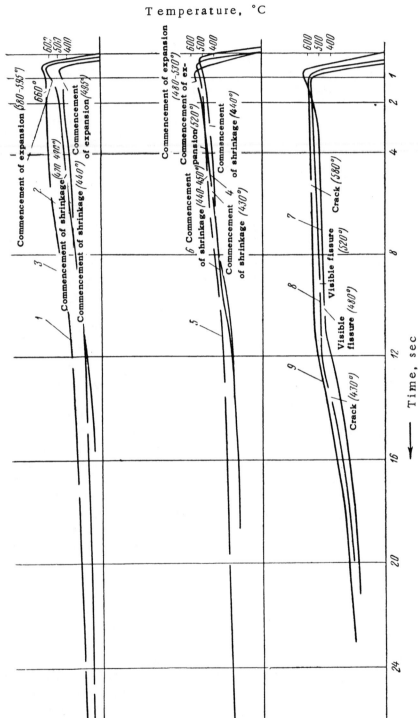

Fig. 105. Solidification and shrinkage oscillograms of aluminum and its alloys: 1) Al; 2) Al+15% Mg; 3) Al+25% Mg; 4) Zn+15% Al; 5) Zn+23% Al; 6) Zn+25% Al; 7) Al+40% Cu; 8) Al+45% Cu; 9) Al+50% Cu.

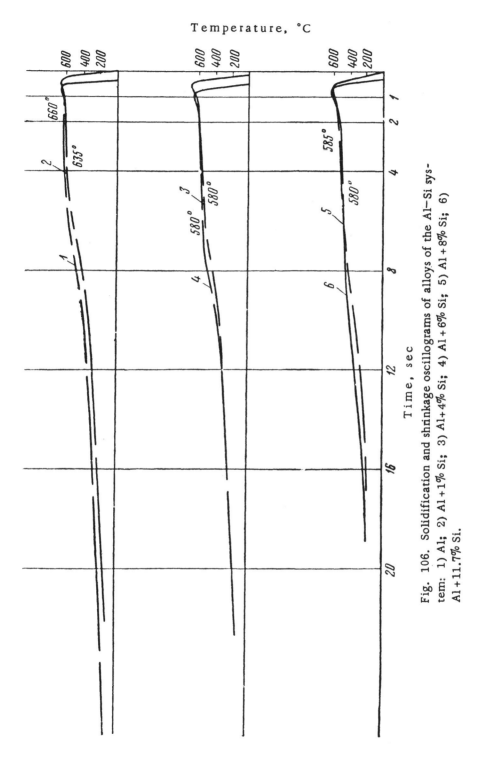

Temperature, °C

Time, sec

Fig. 106. Solidification and shrinkage oscillograms of alloys of the Al−Si system: 1) Al; 2) Al+1% Si; 3) Al+4% Si; 4) Al+6% Si; 5) Al+8% Si; 6) Al+11.7% Si.

The oscillograms also show that shrinkage begins at temperatures corresponding (according to the constitutional diagram) to the solidification range, indicating that in the crystallizing specimen at these temperatures a sufficiently strong crust of primarily precipitated crystals is present, which commences to show the property of the solid casting, i.e., linear shrinkage. A very interesting fact observed in making the oscillograph records is that the rate of shrinkage of such alloys at the very beginning (in the crystallization range) for equal cooling conditions is much higher than for the pure metals and eutectics. Characteristic data are given in Table 30.

TABLE 30. Variation in Temperature and Shrinkage Rate During Crystallization and Cooling.

Alloys	Temperature of commencement of shrinkage, °C	Time of commencement of shrinkage from commencement of crystalization, sec	Temperature and time for which shrinkage is increased by 0.192%		Mean shrinkage rate, %/sec
			°C	sec	
Al pure	660	0.5	660—625	4.1	0.048
			625—570	1.6	0.12
			570—510	1.8	0.106
			510—450	3.0	0.064
			450—385	4.8	0.039
			385—335	7.0	0.027
Al + 2.5% Cu	580	2.1	580—560	1,4	0.137
			580—525	0.83	0.231
			525—465	1.3	0.138
			465—405	2.3	0,083
			405—360	4.0	0.048
			360—310	7.5	0.025
Al + 33% Cu	548	0.5	548—525	3.66	0.052
			525—485	2.0	0.096
			485—420	1.66	0.115
			420—350	2.73	0.07
			350—295	5.2	0.037
Al + 1.0% Si	635	2.3	635—580	1.9	0.101
			580—545	1.16	1.165
			545—465	1.83	0.104
			465—374	2.6	0.073

The observed facts confirm that linear shrinkage of some alloys actually commences at temperatures corresponding to the crystallization range ("effective" range); in this range and near it, shrinkage increases very rapidly, more rapidly than in a metal crystallizing at constant temperature. It is possible that the rapid increase in shrinkage of alloys in this temperature region is due to the abnormal expansion of metals near the melting point when heated, as found by V. F. Gachkovskii and P. T. Strelkov [194] in their investigations. It is also possible that it is due to the liberation of gas on solidification. It is obvious that these facts should be taken into consideration when we speak of the production of crystallization cracks in conditions of hindered shrinkage, when the metal is passing through the "effective" crystallization range, in which solid and liquid phases are present, the solid phase being subjected to an intense contraction-shrinkage.

Summarizing what has been said in the foregoing on shrinkage phenomena in metals and alloys, the following conclusions may be drawn:

1. Shrinkage phenomena occurring in the solidification and cooling of metals and alloys play an important part in the production of castings and ingots; in certain cases, these phenomena make it impossible to produce sound

castings (for example, in the production of ingots from certain alloys by continuous casting), due to the formation of shrinkage cracks during solidification.

2. The value of the shrinkage of pure metals is determined simply by calculation from the coefficients of expansion and the melting point, unless there is preshrinkage expansion.

3. The shrinkage of alloys varies regularly with variation in their composition; the value of the shrinkage of alloys is related to their position in the constitutional diagram and can also be determined from the above-mentioned thermal characteristics.

4. The regular character of the "composition — shrinkage" curves obtained for alloys compared with the form of the constitutional diagrams shows that for a number of alloys solidifying in a temperature range, linear shrinkage is produced before final solidification while residual liquid is still present in the casting skeleton which has been formed.

5. The form of "composition — shrinkage" curves makes it possible to indicate the concentration of alloys in which the occurrence of crystallization cracks is possible; the occurrence of cracks depends on both the temperature (vertical) and concentration (horizontal) solidification range.

Formation of Stresses and Cracks Due to Shrinkage of Alloys

To ascertain the relationships between the variation in shrinkage, the character of the solidification of alloys of different compositions and the formation of cracks in castings during solidification in the crystallization range, it is expedient to consider the published theoretical and experimental data on the formation of crystallization cracks in cast specimens of different alloys for the case of hindered shrinkage.

We shall not discuss here problems connected with the production of thermal stresses in castings and the mechanical properties of metals at high temperatures (after solidification), which is very important when examining the causes of the formation of hot and cold cracks in the cooling process of a completely solidified casting. These questions have been discussed in detail in the current literature and textbooks on foundry practice [7, 8, 80, 170, 171, 178, 190 and others]. They describe the strains set up in cooling castings under the action of internal forces resulting from the irregular development of shrinkage and its retardation in different layers of the casting and also as the result of hindered shrinkage due to the resistance of the mold. For estimating the magnitude of the deformations resulting from these stresses, diagrams have been constructed showing the distribution of the temperatures and the compressive and tensile stresses for different zones of cooling castings. In the majority of such investigations, however, castings are regarded as completely solidified bodies with their capacity to accumulate stresses, to be deformed and to be destroyed, when the stresses begin to exceed the limits of strength or plastic characteristics peculiar to a given alloy in the solid state at high temperatures. Since the true strength characteristics of alloys at temperatures close to the solidus is unknown for many alloys, it is not infrequently assumed without evidence that hot cracks are produced by thermal stresses exceeding the permissible. It is at the same time considered that hot cracks are formed in the range of temperatures in which the metal has no elastic properties, while cold cracks appear at temperatures at which these properties predominate.

Yu. A. Nekhendzi [80] points out that the occurrence of hot cracks in low-carbon steel is connected with the low plasticity of the steel at high temperatures and considers evidence of this to be that traces of plastic deformation are never found at the sites of hot cracks. It is here necessary to bear in mind, however, that traces of deformation may be easily destroyed by the annealing of the castings while they are cooling in the mold in a high-temperature zone.

The possibility of the production of cracks in steel at temperatures close to the solidification point is determined according to B. B. Gulyaev by the strength, elasticity, and plasticity of the solidified steel. His work has shown that the risk of the cracking in ingots increases sharply with increase in their dimensions [55].

Opinions are also held to the effect that many alloys at high temperatures, when shrinkage begins to develop, possess such high plasticity that no residual stresses are set up in the metal [195].

K. V. Peredel'skii [157] considers that the susceptibility of solid-solution alloys to hot-shortness increases with increase in concentration of the solid solution.

A. G. Spasskii [7] assumes that hot cracks may be produced both in the solidification period of castings and also during cooling in the solid state. He also states that in certain conditions, hot cracks formed during the solidification period, may be "healed" by residual eutectic liquid. Yu. A. Nekhendzi accepts the possibility of the "healing" of superficial hot cracks in steel castings on condition of high fluidity of the steel; other investigators have refuted such a possibility [196]. A. A. Ryzhikov considers that hot cracks may be formed in the "presolidus" period [197].

The occurrence of hot cracks in plain carbon steels is often attributed to the cooling period of the castings in the range 1250 - 1450°C [55, 80, 172]. Nevertheless, despite the obvious proximity and even coincidence of this dangerous temperature range with the termination of the solidification of some varieties of steel, it is usually considered that the principal steps for combating hot cracks ought to be the provision of pliability in the molds and their correct filling and cooling, as well as an increase in the mechanical properties of the metal at high temperatures, especially the plasticity, the increase in which reduces hot cracking more effectively than increasing the strength [172]. The view sometimes expressed to the effect that hot cracks in alloys are due to high linear shrinkage [198] does not, in our opinion, accord with the facts.

It is here appropriate to quote data on the strength of steel and certain alloys during solidification or at temperatures close to the solidus. According to A. P. Pronov [196], the strength of steel containing 0.2 - 0.3% carbon in the solidification period in a metal mold varies in the limits of 0.5 - 1.0 kg/mm^2, which is three times the value for solidification in a sand mold.

According to the results of British investigators quoted in [80], steel with 0.04 - 0.1% C at a temperature of the outer crust of 1405°C (solidified in a sand mold) ruptured at 0.16 - 0.19 kg/mm^2. B. B. Gulyaev [55] assumes the strength of steel during solidification to be 0.1 kg/mm^2, i.e., one-fifth to one-tenth of that found by A. P. Pronov. D. K. Butakov determined the strength of steel with 0.11 and 0.33% C during solidification and found it to be 0.4 - 0.8 kg/mm^2 [199]. The relative strength of aluminum during crystallization was found to be 0.058 kg/mm^2. It drops to 0.002 kg/mm^2 for alloys of aluminum with a small quantity of copper or silicon, and increases again on further increase in alloying metal content [178].

It appears to us that the occurrence and elimination of hot cracks in castings in widely differing alloys are connected not so much with the mechanical properties of the metal at high temperatures as with the character of the solidification process and with the mechanism of the formation and propagation of shrinkage during the solidification process.

As already mentioned, for indicating the various casting properties of alloys, the number of currently available diagrams of "composition – property" (in the liquid and liquid-solid states) are very few, negligibly few compared with those constructed for the study of the properties of alloys in the solid state. The construction of such diagrams, permitting the process of crystallization as a function of time to be examined in conjunction with shrinkage, segregation, liberation of gas and so forth is a pressing metallurgical problem.

The lack of work in this direction is partly explained by the fact that, for example, the methods of combating hot-shortness in alloys have been mainly directed toward finding means of increasing the strength and plasticity of alloys at high temperatures in the solid state, while insufficient attention has been paid to the analysis of the solidification processes of alloys in relation to their position in the constitutional diagrams. The failure to appreciate this line of investigation of casting processes is even evident in the fact that so far practically no work has been done to show the quantitative "composition – hot-shortness" relationship for alloys of the Fe−C system, the most important one for metallurgists.

One of the first of the investigations of which we are aware on the establishment of such a relationship in alloys was the work of the Hungarian investigator J. Verö [200]. He showed that the cracking of castings of Al−Si alloys during solidification in conditions of hindered shrinkage occurs in alloys in a definite concentration range, especially with a content of 1.6% Si, i.e., in the presence of 12 - 13% eutectic liquid toward the end of crystallization of the alloy. Taking into account the unfavorable form of distribution of this liquid in the structure of the casting in course of formation, J. Verö regards the main cause of cracking to be the presence of residual liquid phase during solidification. If the silicon content of the alloy is increased, the amount of residual liquid toward the end of crystallization increases and the tendency of the alloys to hot-shortness is considerably diminished.

In the work of the Institute of Metallurgy, carried out under the direction of Academician A. A. Bochvar in the field of nonferrous and light alloys and Academician N. T. Gudtsov in the field of ferrous alloys, the theory of the hot-shortness of alloys during solidification has been the subject of considerable development as part of the general theory of the crystallization of alloys. Thus, in the example of the binary alloys of aluminum with copper and silicon, it has been shown that the stresses at which cracking occurs during solidification vary as a function of the composition and have a minimum, corresponding to a definite alloy composition.

The maximum stresses on solidification are resisted by aluminum and high-eutectic alloys. For a crack to be produced in the solidification of pure aluminum, it is necessary to produce hindered shrinkage relatively equal to 0.058 kg/mm^2; for hot-short alloys, this stress is 1/20 - 1/30 of that value [178].

These conclusions have been confirmed in work by V. I. Dobatkin [170], V. A. Livanov [201], I. N. Fridlyander [202] and others on the production of aluminum alloy ingots by continuous casting, and also in the development of welding processes [203 - 205].

A. P. Pronov [196] in an investigation of shrinkage phenomena in carbon steels, examined the cause of hot cracking in the light of the competing factors strength and shrinkage at the crystallization temperatures. According to his observations, shrinkage of 0.3% C steel begins to develop intensely at the solidus temperature (1450°C), which is critical in regard to cracking.

D. K. Butakov, using the methods of electrical conductivity and thermal analysis, found that cracking began at 1360°C and 1325°C in the solidification of steel with 0.11 and 0.33% C [199].

It should be remarked that some authors ignore the effect of impurities, resulting in the occurrence of fusible components on the dendrite boundaries, the existence of which cannot be recorded by thermal analysis. It should also be pointed out that in our discussions of questions relating to the connection between structure and the character of crystallization, normal cases of the solidification of alloys in ingot molds and foundry molds are meant, without considering modification phenomena.

For normal cases of solidification of alloys it is admitted that for a given alloy composition, the rate of solidification is one of the factors which determine the structure and mechanical properties of the casting.

Solidification of Alloys and the Formation of Crystallization Cracks

High rates of solidification are opposed by a number of factors connected with the construction of the machines and molds for pouring the metal and producing ingots and castings, as well as with the properties of the alloys themselves, mainly the thermal properties, thermal conductivity, specific heat, heat of crystallization, and also with the occurrence of shrinkage cracks in the castings. The thermal factors are usually considered in the analysis of the solidification processes in experimental and theoretical work [11, 26, 55, 56, 76, 80, 189, 204, 206 and others]. At the same time, factors such as the variation in the heat of primary and secondary crystallization of alloys of different compositions during solidification are often disregarded. In estimating solidification rate and the character of the structures in course of formation, it must be remembered that for given conditions of heat transfer from metal to mold, the rate of cooling and solidification will be higher for a metal of relatively high thermal conductivity and having a lesser heat of crystallization. Assuming for individual alloys which are not strongly alloyed, the heat of crystallization is approximately equal to that of the basic component of the alloy (for which assumption there is adequate justification [146]), it is to be expected that the cooling rates of different alloys, for the same conditions of heat transfer, will be proportional to their thermal conductivity. The thermal conductivity of many pure metals (Al, Zn, Pb and others) in the liquid state is less than in the solid state by approximately 30 - 40% [179]. Therefore, the cooling rate of metals in the liquid and liquid-solid states, for the same conditions of heat transfer, will be much less than in the solid state.

Heat transfer, as a diffusion process, is due to the thermal vibrations of the atoms and the motion of the electrons, which are strongly impeded in solid solutions compared with the pure metals, due to distortion of the atomic crystal lattice of solid solutions by the presence of foreign atoms. Thus, aluminum bronze with 6 - 8% Al has a thermal conductivity one-quarter that of copper at the normal temperature; the thermal conductivity of saturated solutions of copper and magnesium in aluminum is from one-half to one-quarter that of aluminum, and so forth. The character of the variation of thermal conductivity with variation in composition is shown in Fig. 62 for a number of alloys. The thermal conductivity of industrial alloys is reduced still further as they become more complex in composition, and as components are introduced which enter into solid solution (alloys of duralumin type, special bronzes, etc.).

Of course, the solidification of these alloys having a considerable crystallization range, i.e., lengthy coexistence of the solid and liquid phases, will be much slower than the solidification of pure metals or alloys not possessing mutual solubility.

In fact, experimental work on the determination of the time of complete solidification of metals and alloys shows that pure metals solidify much more rapidly than some alloys based on these metals. Reference has already been made above to one of the investigations of Academician A. A. Bochvar [74] showing that for identical thermal conditions, the time for complete crystallization of pure aluminum is 0.6 - 0.5 that of the same quantity of aluminum alloys with 5 and 12% Cu, solidifying in the temperature range 100 - 80°C. Similar results have been obtained by British investigators [75, 76] in the thermal analysis of castings of identical form and dimensions of pure aluminum, copper and alloys based on these metals, solidifying in identical conditions, and also by B. B. Gulyaev and O. N. Magnitskii [237] in an investigation of the influence of the composition of alloys on the kinetics of solidification of castings.

The results of our experiments (Table 31) also show definitely that the addition of relatively small amounts of silicon and particularly magnesium to aluminum results in an appreciable increase in the time for complete solidification.

It is evident that with approximately the same quantity of heat liberated in the crystallization of pure aluminum and in that of the primary and secondary phases in the case of the alloys, the increase in the solidification time of the latter is due to the presence of liquid phase of low thermal conductivity in the crystallizing alloy, and also to the reduction in thermal conductivity of the aluminum solid solutions containing silicon or magnesium.

TABLE 31. Solidification Time of Aluminum and Its Silicon and Magnesium Binary Alloys

Alloys	Solidification time, min	Solidification range, °C	Alloys	Solidification time, min	Solidification range, °C
Al pure	26-27	–	Al + 2.5% Mg	50	200*
Al + 2% Si	28	80	Al + 5.0% Mg	49	180*
Al + 5% Si	30	45	Al + 7.5% Mg	45	160
Al + 8% Si	37	25	Al + 10% Mg	39	150
Al + 11.7% Si	25	–	Al + 15% Mg	45	120
Al + 15% Si	38	70			

*For equilibrium conditions of solidification

The observed facts lead to definite concepts regarding the relationship of the character of solidification of alloys to the development of shrinkage and the occurrence of crystallization cracks in hindered shrinkage. Analysis of the cooling curves of Al-Si and Al-Mg alloys, given in Figs. 75 and 76 shows that for alloys with 2 - 2.5% of alloying additions, the heat of crystallization of the bulk of the alloy keeps the temperature practically constant and slightly below the solidification temperature of the pure aluminum*. The difference is that for the alloys, an appreciable drop in temperature to the solidus point is observed only at the moment of separation of the bulk of the primary crystals, when a continuous casting skeleton has already been formed. This temperature drop is followed by an intense development of shrinkage which, if there is any hindrance, causes cracks to appear at the joints of the dendrites, since the latter are separated by residues of still unsolidified liquid. It should be borne in mind that, for foundry practice, when examining shrinkage phenomena and the development of shrinkage, it is important to know the temperature distribution in the various zones of the casting, and to have data on the course of solidification and the quantitative proportion of solid and liquid phases.

Experimental work [75, 76] has shown that in the solidification of aluminum in sand molds, the temperature of the external and internal parts of the casting (ingot 125 mm in diameter) does not differ by more than 10°C at the commencement and by 25°C at the end of solidification. For the alloy Al + 4% Cu, the difference in the temperatures at the surface and center does not exceed 10 - 15°C throughout the entire 50-min period of its solidification. Bearing in mind that the solidification range of this and other alloys attains 100°C or more, it must be considered that solid and liquid phases coexist in the solidification process in all sections of such castings. The relative content of liquid phase in the surface layer of the ingot will of course be less than in the interior, but it may be retained there practically up to complete solidification of the entire casting. At that moment, when the skeleton of the casting has been formed and shrinkage of the alloy commences, there is a real danger that crystallization cracks will be formed. The greater the difference in temperature at the center and surface, the greater is the amount of shrinkage and consequently also the greater is the danger of cracking in hindered shrinkage (in cases of the solidification of alloys in a temperature range), in view of the presence of layers of liquid phase between the dendrites in the outer part of the casting. Such a picture of the formation of external cracks in the cooling of steel blocks was described by D. K. Chernov: "The slightest irregularities on the walls of the mold, impeding the contraction of the cooling skin are sufficient to overcome the bond between the prisms" [10, p. 179] (by contraction is understood shrinkage, and by prisms, dendrites. –A. K.).

Since for pure metals and eutectics, the transition to the solid state occurs with a "continuous front," i.e., without the formation of a skeleton in the effective crystallization range, and consequently without layers in the solidifying parts of castings, hindered shrinkage of the external part ought not to cause the dendrites to become separated from each other; hot-shortness in castings of such alloys is in fact not observed.

It is evident that in rapid cooling, the difference in temperature between the external and internal volumes of metal will be greater than the values given above for castings in sand molds.

---

*The curves in Fig. 75 have been separated slightly in the vertical direction for the sake of clarity.

Fig. 107. Microstructure of commercial aluminum (× 21).

Fig. 108. Microstructure of the alloy Al + 0.12% Fe (× 40).

For any cooling rate of an actual casting of an alloy solidifying in a temperature range, accomplished by means of external factors (cooling in sand or metal molds, cooling in water-cooled molds or directly with water), however, the external part of the casting passes through a stage of two-phase state. The duration of this stage will be longer, the lower is the thermal conductivity of the alloy (in particular, the thermal conductivity of the liquid part of the alloy) and also the greater is the heat of crystallization and the wider is the solidification range.

However small the shrinkage of the solidifying parts of a casting may be, the presence of liquid on the boundaries of the dendrites may result in the separation of the latter, i.e., it may produce hot cracks in the case of hindered shrinkage of the casting, weld or the like. These cracks, as may be seen in the photomicrographs in Figs. 107 and 108, faithfully reproduce the outlines of the boundaries of adjacent dendrites, with their projections and recesses, without any appreciable traces of deformation in the adjacent portions of metal. It is evident that they have been formed in the presence of a liquid film between the dendrites, when no appreciable stresses are required for the separation of the latter. The structural microdefect started in this way develops as a visible crack in the subsequent action of shrinkage during cooling in the effective crystallization range and afterwards, from the solidus temperature to room temperature, i.e., when the principal amount of shrinkage takes place. This is shown by an analysis of the above-mentioned structures, since normal, perfectly sound boundaries pass alongside the defective dendrite boundaries. If a certain quantity of residual liquid were left at the end of crystallization at the position of a defective microvolume, the microscopic crack would be filled with this liquid and no further development would occur.

Before proceeding to discuss the evidence in support of the possibility of such "healing" of crystallization cracks, we shall consider in some detail the experimental data which have been obtained on the character of "composition – shrinkage" curves in relation to the constitutional diagrams, and also the pattern of the deformations produced in metals and alloys during their solidification in conditions of impeded shrinkage.

Dependence of Hot-Shortness of Alloys on Their Composition and the Form of the Constitutional Diagrams

Two simple methods have been used for ascertaining the relationships governing the susceptibility of alloys to hot-shortness during solidification as a function of their composition and the form of the constitutional diagrams.

One of them was to determine the "critical cross section" of cast specimens of the alloys for which no hot cracks appeared in conditions of completely hindered shrinkage [207]. The arrangement of the experiments in using this method was as follows: specimens of different cross section but of equal length were cast from a melt of the alloy of given composition in the same thermal conditions. The specimens in which a crack has formed after solidification in conditions of completely hindered shrinkage were then selected. The magnitude of the "critical cross section" of the specimen (mean of the section with the crack and the nearest section in which there was still no crack) determined the relative susceptibility of the given alloy to hot-shortness during solidification. After testing a number of alloys of different compositions in this way, it is possible to construct a "composition – hot-shortness" diagram.

The second method ("critical composition" method) was to cast specimens of the same size from alloys of different compositions in identical conditions with completely hindered shrinkage. The occurrence or absence of cracks in specimens of a given alloy also enabled a "composition − hot-shortness" diagram to be plotted and a relationship to be established between hot-shortness and the constitutional diagram.

Figures 94, 109, 110, and 111 show diagrams illustrating the results obtained by these methods for the binary alloys Al−Si, Al−Sn, Cu−Si, Cu−Al, Cu−Sn. The form of the hot-shortness curves clearly shows that pure metals and alloys, which are relatively rich in liquid component toward the end of crystallization, are not susceptible to hot-

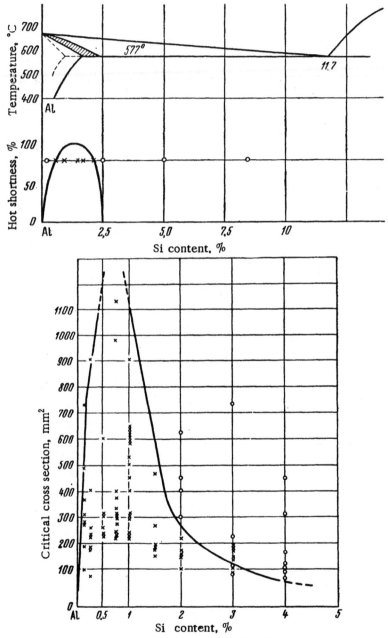

Fig. 109. Dependence of hot-shortness of Al−Si alloys on composition (x − cracks; O − no cracks): a) "Critical composition" method; b) "critical cross section" method.

shortness during solidification. The absence of cracks in pure metals is due to the character of the crystallization taking place at constant temperature, when the hard skin or crust of the casting which forms is continuous and is not divided up by layers of residual liquid, as is found in the case of hot-short alloys. Since the thermal conductivity of pure metals is high, the growth of the skin occurs at a relatively high rate.

The stresses set up by hindered shrinkage result in plastic deformations in the completely solidified metal. The character of these deformations is shown by the photomicrographs of Fig. 112 of the surface of specimens of aluminum and some alloys solidified in conditions of hindered shrinkage. Examination of the surface relief of the specimens under the microscope enables us to assess the behavior of the metal in the microvolumes in which the stresses were concentrated during shrinkage.

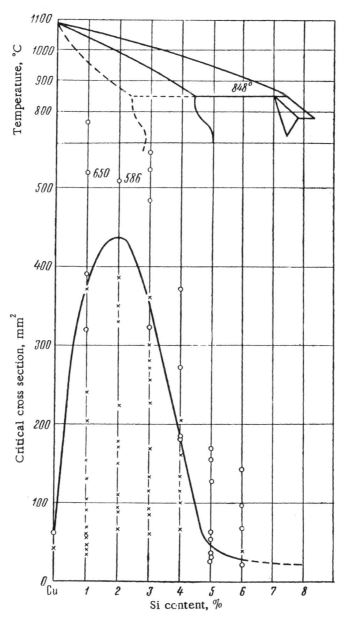

Fig. 110. Dependence of hot brittleness of Cu–Si alloys on composition (same symbols as for Fig. 109).

Under the microscope, it is possible to distinguish the shear lines within the crystals, characteristic of a deformed metal, and also the displacement of the crystal boundaries, which may occur as the result recrystallization of stressed and strained portions of the metal, and also directly from deformation during shrinkage. Most characteristic of all is the presence of slip bands in the microvolumes of cast specimens, resulting from shear strain during shrinkage. In some cases, the slip bands are very narrow, and they are detected with difficulty during microscopic examination of the relatively rough surface of cast specimens; their width, however, is most often 2 - 4 μ (Fig. 112a, b).

From the results obtained it may be concluded that the boundary bonding between the crystals in pure metals during solidification and subsequent cooling is fairly high and may withstand the stresses set up during hindered shrinkage. We have observed similar phenomena in experiments with tin and its alloys.

A different picture is obtained for alloys solidifying in a temperature range and, what is particularly important to keep in mind, passing during solidification through an "effective" range, the physical significance of which has

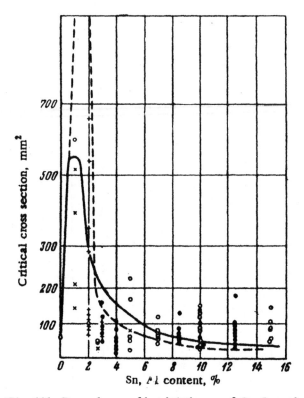

Fig. 111. Dependence of hot brittleness of Cu—Sn and Cu—Al alloys on composition: ✕) Cu—Sn alloys (with cracks); O) Cu—Sn alloys (no cracks); +) Cu—Al alloys (with cracks); ●) Cu—Al alloys (no cracks).

already been pointed out. On the addition of alloying metals, silicon, copper and others, to aluminum, both the internal structure of the dendrites produced and the condition of their boundaries are altered. As a rule, alloying results in a reduction in size of the dendritic cells and in a coarsening of the dendrite boundaries. For a small content of alloying addition, the capacity for displacement of the dendrite boundaries during solidification and cooling is still retained, but on further alloying, the connection between them is destroyed during solidification, in consequence of hindered shrinkage and the presence of a liquid layer, and therefore cracks appear on the surface and in the body of the cast specimens. Microscopic examination of the surface layers shows how the interdendritic spaces are widened as alloying increases, and how the boundaries of the dendrites are disturbed by shrinkage.

In some microvolumes, in addition to "disturbed" dendrite boundaries, it is possible to detect normal boundaries, characteristic of the sound metal (Fig. 113).

In alloys particularly susceptible to hot-shortness, local destructions of the structure become through-going cracks, the width of which after complete cooling is approximately equal to the magnitude of total contraction of the specimen due to shrinkage. Evidently, for such alloys, passage through the "effective" solidification range in the case of hindered shrinkage is accompanied by a general disunion of the primary dendrites, the bond between which is completely destroyed by films of residual liquid alloy.

If polished sections passing through the surface and "depth" cracks are examined under the microscope, it will be seen that the coarse, wide boundary cracks become thin, normal cracks. These sound "terminations" of dendrite boundaries often "rest" on microscopic portions filled with eutectic veins, the presence of which is also characteristic of the adjacent dendrite boundaries in this part of the polished section (Fig. 114).

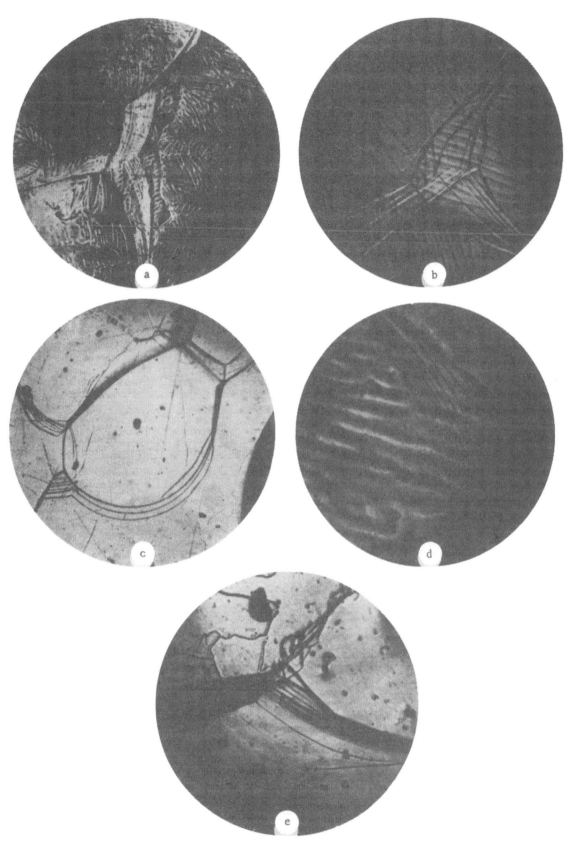

Fig. 112. Traces of deformation due to hindered shrinkage in alloys: a) Alloy
Al + 0.2% Zn (× 21); b) the same (× 210); c) aluminum mark AVOO (× 210);
d) alloy Al + 1% MgZn (× 40); e) alloy Pb + 0.1% Sb (× 210).

It should be emphasized that the quantity of eutectic to be observed in such microvolumes far exceeds the quantity corresponding to the mean composition of the given alloy; this is evidently due to the flow of residual liquid during the solidification process into the "voids" or shrinkage cracks. We here come to the mechanism of the elimination of crystallization cracks by their self-healing by residues of liquid parent solution in the solidification process. The possibility of such healing has been indicated in previous work [152, 178], but has sometimes been doubted by other investigators [196].

Fig. 113. Destruction of dendrite boundaries on shrinkage of the alloy Al + 0.05% Fe (× 90).

A typical structure of a microvolume with healed shrinkage cracks is shown in Fig. 115 for the case of hot-short aluminum-silicon alloys; this structure corresponds to an alloy with 1.5% Si, in the range of hot-short alloys (see Fig. 109). In this case, healing of the cracks occurs through liquid eutectic filling either the very thin microscopic cracks, the width of which does not exceed 5 - 10μ, or the "commencements" of wider cracks starting from the surface of the specimen. In one way or the other, only local microscopic accumulations of liquid alloy have flowed into the free interdendritic spaces from the adjacent regions and "accidentally" healed the crystallization cracks. The residual liquid is not sufficient for healing a large number of rapidly produced cracks, and the hindered shrinkage of the crystallizing specimen results in its fracture across the entire section. Such alloys have maximum susceptibility to cracking on solidification, indicated by a maximum on the hot-shortness curves. The descending part of these curves (in proportion to the further addition of alloying component) corresponds to a state of the crystallizing alloys where the quantity of residual liquid is sufficient to fill the voids between the growing dendrites and to heal the cracks produced between them during passage through the "effective" crystallization range. The structure of a microvolume of such alloys with "healed" cracks is shown in Fig. 115, where a previously formed, relatively wide crack has been filled with eutectic.

Evidently, the further increase of silicon in the alloys results in the disappearance of the dangerous crystallization range and in the accumulation of a large quantity of eutectic liquid, completely permeating the boundaries of the growing primary crystals and filling the crystallization cracks formed. Sometimes, however, even in high alloys, i.e., when their composition is beyond the limits of hot-shortness, microcracks can be seen in the structure of specimens which have solidified under conditions of hindered shrinkage. As a rule, these microcracks are situated in regions connected together by eutectic; such a structure is shown in Fig. 116.

The observed facts confirm with sufficient completeness the casual relationship between the occurrence and elimination of crystallization cracks and the form of the constitutional diagram of alloys. Evidently, for Al–Si and Al–Cu alloys, the "effective" crystallization range and consequently the susceptibility to cracking occur on the introduction of small quantities of silicon or copper (0.1 - 0.3%) and are eliminated at 2 - 2.5% Si or 7.5 - 8.5% Cu, respectively. Inside these limits, the hot-shortness curve of the alloys passes through a maximum, corresponding to the maximum value of the "effective" crystallization range (see Figs. 109 and 111).

Elementary calculations of the relative quantities of solid and liquid phases towards the end of crystallization (on the "lever principle," taking the nonequilibrium position into account) show that for the elimination of crystallization cracks in Al—Cu alloys, there must be present 26% of eutectic liquid towards the end of crystallization, and 18% in

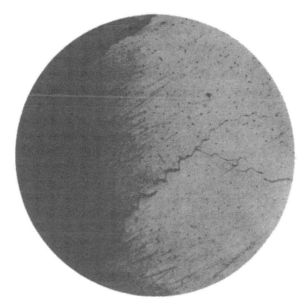

Fig. 114. Crystallization cracks and their "healing" in the alloy Al + 1.5% Si (× 210).

Fig. 115. "Healing" of crystallization cracks formed as result of hindered shrinkage (× 210).

Al—Si alloys. The most unfavorable alloys from the point of view of hot-shortness contain 13 and 9% of residual liquid, respectively. These results are in close agreement with the observations of other investigators [177, 200, 208]. Corresponding data are given in Table 32 for some binary alloys; they are the results of the evaluation of experimental data obtained by the "critical" composition method.

Comparison of "composition – shrinkage" curves (Figs. 93 - 100) and "composition – hot-shortness" curves (Figs. 109 - 111), as well as the data in Tables 29 and 32 for binary alloys of binary systems readily shows the inter-relationship of these two phenomena.

Fig. 116. Microcrack in portions impoverished in eutectic.

The above described operation of "crack healing" does not occur, however, with the same degree of complete-ness in all the defective places of a casting. Circulation of liquid in the intercrystalline capillary passages will de-pend on its quantity and properties (viscosity, surface tension and others) and the magnitude (width and length) of the capillary passages. In addition, as has been shown by recent investigations, the movement of the residual liquid

TABLE 32. Composition of Alloys Susceptible to Hot-Shortness on Solidification in Conditions of Hindered Shrinkage

Alloys	Composition of hot-short alloys, %
Cu–Zn	Up to 2.5 - 3 Zn
Zn–Sn	Up to 3.0 Sn
Al–Fe	0.15 - 0.2 Fe
Al–Zn	Up to 50 Zn (beyond was not investigated)
Al–Mg	Up to 2.0 Mg
Al–Ni	Up to 2.5 Ni

inside the crystallizing mass is intensified by vibratory agitation; a more complete healing of the crystallization cracks is then obtained [22]. These questions are bound up with the extremely interesting problem of the micro-hydromechanics of solidifying alloys, still awaiting elucidation and solution.

It is well known that castings of alloys, solidifying in a temperature range, have in addition to visible cracks other defects associated with shrinkage phenomena, for example, microcracks and macropores. The quantitative variation of the latter in relation to the variation in composition and the form of the constitutional diagrams of some aluminum and copper alloys will be shown in the next chapter. It should here be pointed out that the formation of crystallization cracks during the solidification of alloys is not directly related to the value of linear shrinkage. The highly plastic metals which we have tested and which have different shrinkage values (for example, 1.58% for zinc and 0.63 % for tin) also show hot-shortness when metals are added to them which result in the occurrence of readily fusible constituents. In such cases, we cannot say that the alloys have inadequate plasticity in the solid state, to which hot brittleness could be ascribed. In these alloys, the eutectic constituent is as plastic or even more so than the components of the alloy [209].

Passing to the question of the hot-shortness of ternary and more complex alloys, it should be stated that if, for example, in a binary eutectic hot brittle alloy there are impurities resulting in the occurrence of a readily fusible ternary eutectic, an "effective" solidification range at a definite concentration of the components is also possible.

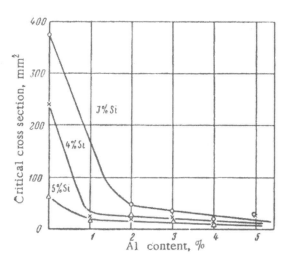

Fig. 117. Dependence of hot-shortness of Cu–Al–Si alloys on their composition.

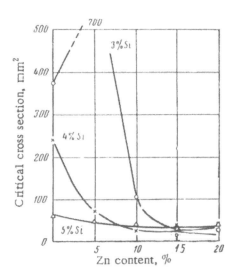

Fig. 118. Dependence of hot-shortness of Cu–Zn–Si alloys on their composition.

As shown by our investigations of alloys of the ternary systems Al–Cu–Si, Al–Mg–Zn, Cu–Si–Zn and others, on which industrial aluminum alloys are based, the causes and mechanism of the production and elimination of crystallization cracks in them are the same as in binary alloys. A characteristic feature of these alloys is the production of hot-shortness on the introduction of comparatively small amounts of alloying elements and the elimination of hot-shortness at a definite "critical" composition. The results of investigations on ternary copper alloys are shown in

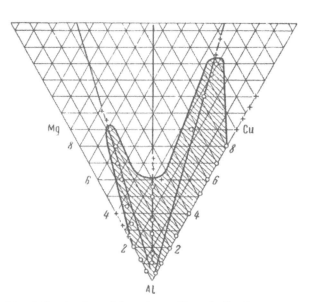

Fig. 119. Region of hot-short alloys in the ternary system Al–Cu–Mg.

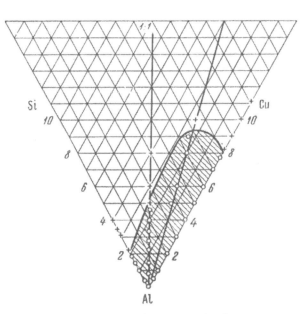

Fig. 120. Region of hot-short alloys in the ternary system Al–Cu–Si.

Figs. 117 and 118 and those on ternary aluminum alloys in Figs. 119 and 120 the shaded parts indicate the composition of alloys susceptible to cracking on solidification).

Thus, the observed facts show that in ternary alloys also there is a possibility of eliminating cracks by a slight variation in composition of the alloys, since the hot-shortness curves pass within fairly narrow limits of concentration of the components in individual sections of the diagram. Thus, for definite conditions of the temperature and speed of pouring and cooling of the specimens, cracks in Al–Cu–Si alloys on the section Cu : Si = 1 : 1 are eliminated on passing from an alloy containing 4.5% (Cu + Si) to an alloy with 5.0% (Cu + Si). A similar relationship is also observed for the other alloys.

Numerous structural analyses of the surface layer of ternary alloys have shown the presence of deformations and defects in microvolumes, as well as their elimination, as described above for binary alloys.

The results we have obtained in the assessment of the hot-shortness of aluminum-based alloys are in good agreement with results of similar work carried out by Soviet investigators in the examination of the hot-shortness of alloys in general [152, 177, 190 and others] and in regard to welding processes [203, 204], and also of British investigators studying ternary aluminum alloys [210 - 213]. One of the diagrams they obtained, showing the hot-shortness of alloys of the Al–Cu–Si system is reproduced in Fig. 121. In contrast to our diagrams which merely outline the region of hot-short alloys, the diagrams of the British investigators also show the total magnitude of the shrinkage cracks. The maximum of this magnitude practically agrees with the maximum on our hot-shortness curves, which expresses the maximum percentage of castings with cracks or the most "critical" cross section of such castings (see Fig. 109). The advantage of our methods is that they give reproducible results in workshop conditions and can easily be adopted by any laboratory. The results obtained enable the susceptibility of any alloys to hot-shortness to be assessed with adequate completeness, and the composition of the alloys and the conditions in regard to temperature and speed of pouring to be corrected.

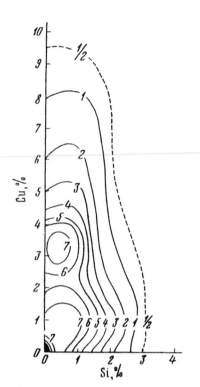

Fig. 121. Region of hot-shortness of alloys of the ternary system Al–Cu–Si (the numbers against the curves denote the total magnitude of shrinkage cracks in inches) [211].

Hot-Shortness in Industrial Alloys

In the theoretical consideration of the question of the shrinkage of alloys and the formation of crystallization cracks which has been given in the foregoing, characteristic examples were cited from foundry practice which showed that the theory developed was entirely applicable to the solution of practical problems. Since the majority of industrial foundry alloys, with the exception for example of cast iron and high-silicon silumin, solidify in some temperature range or other, they all have a tendency to hot-shortness in hindered shrinkage. This tendency, however, is exhibited to the maximum extent only in alloys in which, toward the end of solidification in the "effective" range, there is a relatively small amount of residual liquid melt distributed in the form of thin layers surrounding the growing crystals.

Reverting to already cited work [80, 196, 197, 205 and others], it is pointed out that in these publications, in explaining the causes of the hot-shortness in castings, particular attention is devoted to a clarification of the strength and plastic properties of alloys at high temperatures close to the solidus. There is no doubt that these properties, especially plasticity [190, 214], as well as the temperature-velocity conditions of casting, play a definite part in regard to the reduction of hot-shortness, but the principal source of crystallization cracks must be sought in the two-phase solid-liquid condition of the castings, inevitable for most alloys solidifying in conditions of hindered shrinkage. This point of view is being accepted by a growing number of metallurgists.

Reference will be made to investigations carried out in recent years on hot-shortness in industrial alloys. A number of authors investigating the crystallization process in steel ingots of different compositions, have come to the conclusion that cracks produced as the result of hindered shrinkage are most strongly developed in steel containing 0.2% C having a short solidification range [215]. A. P. Pronov [144] also considered the mechanism of cracking in the solidification of steel from the point of view of the Fe–C constitutional diagram, but with a definite limitation to crystallization kinetics. He points out that in the case of steel with 1% C, practically no hot cracks are formed in conditions of hindered shrinkage, while steel containing 0.15 - 0.30% C is very susceptible to hot-shortness, this applying to both laboratory and industrial conditions. The transactions of a conference on the continuous casting of steel contain papers by V. N. Saveiko [216] and V. G. Gruzin [217], the contents and conclusions of which confirm that the analysis of the causes of hot-shortness in steel should be based on constitutional diagrams and the process of structure formation in the actual casting.

Wide possibilities for the generalization of industrial experience concerning the suppression of hot-shortness in castings and ingots appear from an examination of work on the casting of various alloys. In regard to the production of castings, such generalization is given in previously cited work [26, 75, 80, 197, 208, 218, 219 and others], in regard to ingot production [11, 53, 68, 170, 171, 201, 202] and in regard to welding [203, 204].

Examples from industrial practice are given below, showing the influence of composition and technical casting conditions on hot-shortness in alloys. Thus, in the book by S. G. Glazunov and S. I. Spektorovaya [220] on the determination of the technical properties of light alloys and in the British paper on the standard specification for industrial aluminum alloys [169], it is definitely stated that the hot-shortness of the alloys depends on the "effective" solidification range, i.e., on their position in the equilibrium diagrams. Thus, Al–Si alloys of industrial composition with 6 - 13% Si possess maximum resistance to cracking (according to the three-point system), while Al–Cu alloys with 4 - 5% Cu have minimum resistance. Alloys of the ternary system Al–Cu–Si with 2 - 4% Si and 6 - 8% Cu have an average of one point of hot-shortness [169]. I. F. Kolebnev [221] gives a similar characteristic of industrial aluminum casting alloys belonging to the Al–Si, Al–Cu and other systems. The hot-shortness of magnesium casting alloys is also in good agreement with the "composition — hot-shortness" curves obtained for a number of aluminum alloys. Thus, the hot-shortness of the alloy series ML3, ML5, and ML6 decreases regularly with increase in aluminum content from 3.5 to 10% [142], i.e., with decrease in the effective solidification range and with increase in the residual liquid, assisting the elimination of cracking in the alloys.

In selecting the types of secondary aluminum alloys for foundry purposes [222], it was found that the stress of crack-formation in Al–Zn–Si alloys, with decrease in the zinc from 6 - 10 to 3 - 5.5% and simultaneous increase in the silicon from 1 - 1.5 to 4%, increased by a factor of hundreds, namely, from 0.007 - 0.001 to 0.18 kg/mm^2 or more. Taking into consideration the fact that alloys in this system solidify like alloys of the Al–Si binary system, the results obtained are in full agreement with the "composition — hot-shortness" curves for Al–Si alloys (see Fig. 109), according to which an increase in the silicon content from 0.5 - 2.5 to 3 - 4% makes the alloys resistant to crystallization cracks.

In addition to correcting the composition in counteracting the susceptibility of industrial alloys to hot-shortness, extensive use is made of methods of regulating the conditions of temperature and speed of pouring and cooling of the alloys in the molds. Thus, when casting the magnesium alloy ML5 in permanent molds, hot-shortness of the alloys is eliminated by heating the molds to 400°C, i.e., nearly to the solidus temperature of the alloy (436°C), the pouring temperature being raised to 800°C at the same time [223]. Evidently, this minimizes the negative influence of the "effective" solidification range and increases the above-mentioned possibility of "healing" the crystallization microcracks formed, the character of which is shown in Figs. 114 and 115.

V. A. Livanov [201], in an article on continuous casting, showed that if the quantity of free silicon, left after the iron and silicon impurities have been combined in a ternary compound with aluminum, reaches 0.03 - 0.05%, the number of ingots rejected for surface cracks is considerably increased. If, however, the alloy melt is prepared with iron (0.06 - 0.1%) for combining with the free silicon, cracking of the ingots is eliminated. The influence of the free silicon is explained as being due to the occurrence of a fusible eutectic in the aluminum and the formation of an "effective" crystallization range, resulting in cracking. When there is no free silicon, the fusible eutectic is not formed and solidification occurs at practically constant temperature. Cases of the occurrence of cracks in railcar bearings of alloys of lead with sodium and potassium, described in the literature [59], also confirm the existence of "critical" compositions of alloys, capable of solidifying with or without cracks, depending on the variation in content of the components.

Much evidence of the influence of a "critical" composition of alloys on hot-shortness may also be found in published work on the investigation of the quality of steel castings. Thus, it has been shown that when manganese steel contains more than 0.09% phosphorus, the susceptibility to the occurrence of hot cracks in the castings is sharply increased [263]. An important part is played in this by the carbon content of the steel, which should not exceed 1.3%; if the carbon content is higher, the number of hot cracks also greatly increases. Textbooks on steel castings (for example, [80]) give many examples of the critical composition of steel in regard to phosphorus, sulfur, carbon and their relationships in causing the development of hot cracks. Despite the fact that the occurrence of hot cracking is explained in these publications as being due to the existence of stresses in the solidified castings, it is certain the principal cause of cracking is the destruction of the continuity of structure in the solidification process. These structural defects develop into visible cracks on subsequent cooling of the castings.

Statements to the effect that one of the causes of the occurrence of filaments of extra-axial heterogeneity ("whiskers") in steel ingots may be the formation of cracks in the solid-liquid state and the filling of these cracks by a fusible constituent [264] are also not uncommon.

The foregoing results show the possibility of eliminating cracks in castings by varying the composition of the alloys, such variation being so insignificant for the individual alloys as to cause no departure from the standardized composition. Such an approach is simpler than, for example, lengthy searches for alloying additions for increasing the strength or plastic properties of the alloys at high temperatures near the solidus.

The results of the investigations also show that the occurrence of crystallization cracks and their healing during solidification is much more frequent than would appear from the technical literature. There is reason to believe that the accumulation of fusible constituents (sulfides, phosphides and other constituents), observed in the form of laminae, pockets and the like in the cast structure of different alloys is not so much the result of segregation alone as the consequence of the filling of crystallization cracks and other shrinkage defects by residual liquid.

The possible method which has been indicated for the elimination of hot-shortness in alloys by varying the composition is an important addition to already known methods for counteracting this phenomenon. These methods are based on varying the conditions of the temperature and speed of pouring of the metal and its cooling.

Chapter VII

THE CHARACTER AND DISTRIBUTION OF SHRINKAGE CAVITIES

The previous chapter dealt with the relationships in the variation of external (linear) shrinkage of metals and alloys occupying different positions in the constitutional diagrams and the possibility of calculating the amount of shrinkage from the thermal characteristics of the metal. The causes of the formation of internal defects, crystallization cracks, in alloys of a definite composition were also pointed out and the measures for eliminating these defects were indicated.

The present chapter examines questions concerning the systematic production and variation in form of another kind of shrinkage defect, i.e., shrinkage cavities (pipes), regions of shrinkage porosity and scattered shrinkage porosity, resulting from the differences in the way in which alloys of different compositions crystallize. These casting defects are an inevitable evil and result in loss of labor, due to rejection of castings and the loss in yield of useful metal, caused for instance by the removal of discard in which the shrinkage cavities are concentrated. The amount of discard in the production of steel castings is 40 - 70% in the best case for castings weighing from 3 to 10 tons; it increases with decrease in the weight of the casting [224]. The loss of metal is substantial, even in the case of ingots; for example, 12 to 15% of the metal is lost in the discard when casting 7-ton rail steel ingots [225].

Foundry practice incurs considerable loss through the existence of internal shrinkage porosity in castings and its unfavorable distribution in the body of castings. The cost of rejects from this form of defect cannot always be estimated in terms of loss of time and loss of some of the metal in using up the scrapped metal in the foundry; to these losses must be added the cost of the mechanical and other forms of processing the casting up to the point where the shrinkage voids are found in it. In some cases, the cost of such processing or the rectification of the defects by far exceeds the cost of the metal itself.

The study of the causes of the different forms of shrinkage defects in castings of alloys of a given composition often relates to individual alloys or to the needs of individual foundries or workshops; the results of such investigations are therefore of a specific character. Our aim has been to study both qualitatively and quantitatively the distribution of shrinkage defects in alloys primarily as a function of composition. After ascertaining the relationships govern-

Fig. 122. Microstructure showing the "survival" of some branches of a dendrite and the suppression of others (× 90).

ing the distribution of the defects in castings of alloys of different compositions, the knowledge acquired can be used as a guide to means for eliminating or minimizing the defects in given alloys and for drawing generalized conclusions.

An examination of the solidification process in alloys shows that for identical solidification conditions, the character and distribution of shrinkage defects will be different in alloys of different compositions.

We have already seen the characteristic difference between the crystallization process of pure metals or eutectics and that of molten alloys solidifying in a temperature range. The former solidify at constant temperature and form a practically continuous front of crystals along the boundary with the liquid phase, since crystals and liquid have the same composition. The presence of impurities or alloying additions, giving rise to a solidification ion range, produces a disturbance in the crystallization front, accompanied by a preferential growth of some dendrites into the liquid and a retardation of others, the growth of which is suppressed by their neighbors. At the same time, the dendrites branch considerably; not only are first order axes retarded in growth, but there is also a considerable increase in the growth of the branches. The branching of the dendrites increases with increase in the amount of alloying element. The characteristic pattern of the irregular growth of the dendrites and their side branches can be seen in Figs. 122 and 123, and also in Figs. 107 and 108.

Fig. 123. Microstructure of Al−Zn alloy.

It is evident that the greater the solidification range and the difference in composition between the solid and liquid phases, the greater will be the development of the dendritic form of crystals (other things being equal). This results in an increase in the extent of the interfaces and consequently in an increase in the probability of breaks in continuity of the cast structure. There can be no doubt that an appreciation of these circumstances facilitates an understanding of the true picture of the production and distribution of shrinkage defects in castings made from alloys of different compositions.

Problems concerning the occurrence of shrinkage voids in a crystallizing metal are more conveniently examined in the following order: 1) Concentrated shrinkage cavity or pipe, 2) regions of shrinkage porosity, and 3) scattered shrinkage porosity.

Concentrated Shrinkage Cavity (Pipe)

A shrinkage cavity or pipe is formed in castings as the result of the reduction in volume of the liquid metal on cooling from the superheat temperature to the solidifying point and during crystallization; the reduction in volume of the metal on cooling in the solid state does not alter the relative volume of the cavity, since with reduction in the absolute volume of the casting, the volume of the cavity is also reduced. The shrinkage cavity is due to the fall in level of the liquid metal relative to the solidified outer skin of the casting, as the result of the thermal contraction of the liquid metal and the reduction in its volume on transition to the solid state. The relative volume of the cavity may be calculated from the following equation proposed by Yu. A. Nekhendzi and N. O. Girshovish [80]:

$$V_p = \alpha_{V_l} (T_{l\,m} - T_s) + \varepsilon_{Vs} - 1.5\alpha_T (T_s - T_{sm}),$$

where α_{V_l}, ε_{Vs} are the coefficients of volumetric contraction in the liquid state and on solidification;

α_T is the coefficient of linear contraction;

T_s is the solidifying point;

$T_{l\,m}$ is the mean temperature of the liquid metal at the moment of solidification;

T_{sm} is the mean temperature of the solid metal at the moment of final solidification of the casting.

It will be seen from this formula that the volume of the shrinkage cavity is mainly composed of the solidification contraction (ε_{Vs}), i.e., the reduction in volume on passing from the liquid to the solid state, which for metals like aluminum and magnesium is 5 - 6% [179], and for steel with 0.1% C is only 2% [80] (see also Table 23). An increase in temperature of the liquid metal increases the cavity, while the decrease in volume during cooling of the outer part of the casting decreases the cavity by raising the level of the liquid in the central portion of the casting.

This formula and others like it are widely used for calculating the volume of the concentrated shrinkage cavity or pipe and for determining the amount of shrinkage head on the casting to ensure the displacement of the cavity from the body of the casting [7, 8, 56, 80, 224 and others].

A number of proposals for reducing the shrinkage cavity or pipe follow from one formula used for its calculation; these include primarily the reduction in pouring temperature of the metal, since the coefficient of contraction of liquid metals is relatively high (for copper it is three times the coefficient of contraction in the solid state, for aluminum and zinc it is one-and-a-half times, and so forth) [179].

Shrinkage cavities concentrated close to the surface of a casting are less dangerous and cause less waste than those which extend along the axis of the casting.

A concentrated shrinkage cavity is often accompanied by regions of porosity, due to insufficient feeding of liquid to the casting and the difficulty which this liquid encounters in penetrating the branching capillary dendritic formations which in the central zone of castings are often more of equiaxial type than columnar.

In this chapter, special attention is paid to scattered porosity, because the literature sheds little light on this problem, and also because it forms the subject of contradictory views.

Shrinkage Porosity (Regions of Porosity and Scattered Porosity)

Shrinkage porosity in castings is the result of shrinkage occurring primarily during the solidification of alloys in a crystallization range, when liquid-solid portions occupy large regions of the casting, most often in those parts where crystallization of the residual liquid phase is being terminated. According to the definition given by D. K. Chernov, "Porosity of the central portion of an ingot is nothing more than an accumulation of more or less developed partial shrinkages" [10].

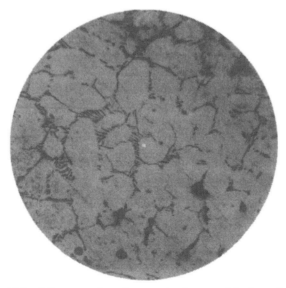

Fig. 124. Microporosity of eutectic-impoverished portions of Al + 6% Si (× 210).

The mechanism of the production of porous portions in castings (Figs. 124 and 125) will be understood from an examination of the crystallization process and from an examination of the structure of the final casting. As a rule, such portions are situated in regions where an equiaxial structure predominates, i.e., where the crystals grow randomly and are not subjected to any directional heat-dissipating effect, promoting the growth of columnar crystals (Figs. 126 and 127).

Some light has been shed on the complex phenomenon of microporosity by the work of N. T. Gudtsov [11, 24]. In describing the crystallization process of a steel ingot, and in particular the zone of disoriented crystals, N. T. Gudtsov states that towards the end of solidification "drops, detached from the liquid, remain in the zone; they freeze independently of the main mass, resulting in the formation of so-called point segregation. This zone has a high concentration of voids in the crystal lattice". In the zone of columnar crystals, however, the concentration of these voids is extremely insignificant. N. T. Gudtsov refutes the widely held opinion that point segregation is due to the segregation of sulfur and phosphorus, since the reduction in mechanical properties and etchability observed in these zones cannot be due to the presence of a high phosphorus content, which actually improves these properties.

In steel ingots and castings, porosity below the pipe and porosity situated along the axis of the ingot are a consequence of the fact that solidification in some portion of the casting was terminated before the liquid metal was able

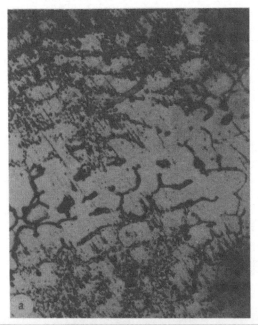

Fig. 125. Porosity in a silumin casting: a) Micropores; b) scattered macroporosity.

to enter the shrinkage voids formed. In nonferrous metal ingots, the mechanism of the formation of shrinkage voids is the same as in steel ingots. In the former, however, especially in light-alloy ingots, shrinkage porosity is often combined with gas porosity; the latter is distinguished from shrinkage porosity by the spherical form of the cavities, since the positive pressure of the gas liberated during the cooling of the liquid and solidifying metal will give the cavities a bubble shape.

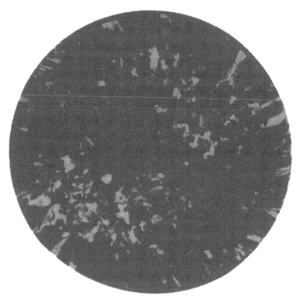

Fig. 126. Macrostructure of a casting of the alloy
Cu +0.1% Al + 0.1% Ti (× 0.8).

Fig. 127. Microstructure of Cu–Si alloy (5% Si) etched with a solution of FeCl₃ in HCl (× 80).

Thus, even a brief examination of the mechanism of the production of shrinkage voids in metal castings indicates that this phenomenon is directly related to the crystallization process and in particular to the formation of crystals of dendritic form. Without entering here into a discussion of hypotheses on the nature of dendrites, their kinetic growth, and the causes of their branching formation (which are given in detail in the references cited above), it may be pointed out that they are mainly attributed to diffusion processes of impurities and alloying additions and their distribution during crystallization. It follows directly from this that there is a difference between the mechanism of the formation of shrinkage defects in pure metal castings and alloy castings.

Unfortunately, in the majority of the investigations on the crystallization of actual alloys and the casting properties of alloys, questions concerning the composition of the alloys and their position in the constitutional diagrams are not always discussed and are sometimes ignored altogether.

Methods have been described in the foregoing for improving the qualities of castings by varying the pouring and solidifying conditions of the metal and the pressure above the solidifying metal. The applicability of a given method to the production of castings from given alloys and the effectiveness of the methods are far from clear. We shall discuss one of them, the method of crystallization under pressure in autoclaves, since in its investigation, the methods of the physicochemical analysis of alloys and the construction of "composition − shrinkage porosity" or "composition − imperviousness" diagrams were used for the first time [26, 84, 86].

By increasing the pressure during the crystallization of alloys to from 5 to 7 atm, liberation of gas from solution is prevented, any gas bubbles which form are compressed and the residual liquid is forced into the rarefied shrinkage voids. Since the crystallization processes and shrinkage distribution are different in alloys occupying different parts of the constitutional diagrams, pressure does not have the same effect in regard to improving the qualities of castings made from different alloys. This is shown by diagrams (Fig. 128) describing the effect of pressure in the crystallization of three types of alloy: 1) pure metals and eutectics, solidifying at constant temperature (Fig. 128a), 2) alloys solidifying in a considerable temperature range (solid solutions) (Fig. 128b), and 3) alloys solidifying in a relatively considerable temperature range, but whose crystallization is completed at constant temperature (eutectic or peritectic alloys) (Fig. 128c).

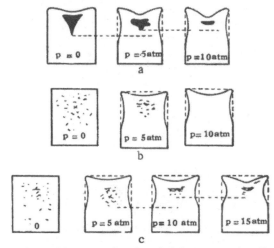

Fig. 128. Diagrams showing shrinkage porosity distribution as a function of crystallization range and external pressure [84].

These porosity distribution diagrams show that alloys of the second group, solidifying with the formation of intensely branching dendrites and, consequently, with the formation of porosity scattered throughout the entire volume of the casting, are consolidated by reduction of the external dimensions of the castings; they form practically no concentrated activity or pipe.

Alloys of the third group are also consolidated by the action of pressure, but in order to obtain good density, it is necessary to have a high pressure and a considerable feed head to "draw" into the latter the pipe produced in the castings, because they do not yield to reduction in their external dimensions.

Alloys of the first group (either saturated with gas or free from gas) produce a concentrated shrinkage cavity (pipe) in agreement with the solidification of metals at constant temperature. A characteristic feature of their crystallization is the rapid formation of a strong outer skin and the continuous advance of the crystallization front from the walls of the mold accompanied by a gradual fall in level of the liquid, which has the same temperature as the solid crystals. If for some reason, the supply of liquid feeding the cavity is interrupted, shrinking "recedes" into the interior of the casting but it will still tend to assume the form of a concentrated cavity or pipe. It will be found to lose its concentrated character when the continuously moving front of columnar crystals is disturbed and gives place to equiaxial crystals formed in the liquid melt. The flow of liquid to the branching boundaries of these crystals will become difficult, due to the reduction in metallostatic pressure as the result of the separation of the column of liquid into a number of individual proportions, and as the result also of the increase in length of the capillary passages along the crystal boundaries.

The subdivision of shrinkage into external shrinkage (for example, the formation of a gap between the wall of the mold and the metal) and internal shrinkage appearing in the form of a concentrated cavity or in the development of porous regions and scattered porosity has proved fruitful in providing an understanding of the true picture of the formation and distribution of the various forms of shrinkage voids in castings. This has been assisted by work carried out by K. I. Akimova [226] with the object of ascertaining the relationship between shrinkage phenomena and the composition of alloys. In this work, alloys of the Al—Si and Al—Cu systems were studied, castings being made from these alloys in the form of spheres solidified in autoclaves under a pressure of 7 atm to prevent any effect of gas liberation. It was found that pure aluminum and eutectic alloys produce a shrinkage cavity or pipe, while alloys solidifying in a temperature range produce external shrinkage, local shrinkage porosity and scattered shrinkage porosity (without a concentrated shrinkage cavity which begins to appear in alloys approaching the eutectic composition). The difference in density between Al—Si alloys as determined experimentally and as calculated was found to be different for alloys of different concentrations. This difference was small for alloys having a large external shrinkage and very large for alloys having internal shrinkage in the form of porous regions and scattered porosity. The results of this work have been applied successfully to the solution of industrial problems [85]. Merely by replacing an aluminum-copper alloy by an aluminum-silicon alloy, found to have good imperviousness in laboratory tests [86], a considerable wastage of castings was remedied. Subsequently, relationships were found between the distribution of shrinkage voids in castings of copper-based alloys and magnesium-based alloys of different compositions.

Dependence of Shrinkage Porosity (Imperviousness) of Copper-Based Alloys on Their Composition and the Form of the Constitutional Diagrams

Castings with porosity scattered throughout the entire volume cannot be perfectly impervious; the degree of this imperviousness may be measured by the specific gravity method or by hydraulic tests. The latter method has also been used for discovering the quantitative relationship between alloy composition and scattered porosity, first in work by A. A. Bochvar and Z. A. Sviderskaya on aluminum alloys [86] and then by the present writer on a number of copper alloys [227].

The construction of "composition — imperviousness" curves was based on the results of the testing of cast specimens of identical dimensions, produced in identical conditions (specific conditions of melting, superheat, etc.). The specimens were taken from the bottom half of flat castings, and were subjected to hydraulic test at a pressure of 70

TABLE 33. Hydraulic Resistance of Copper Alloys

Composition of alloys, %		Crystalliza-tion range, °C	"Critical" thick-ness (occurrence of leakage), mm	Composition of alloys, %		Crystalliza-tion range, °C	"Critical" thick-ness (occurrence of leakage), mm
Cu	Sn			Cu	Al		
100	—	—	1.0 (no leakage)	100	—	—	1.0 (no leakage)
98.5	1.5	35	1.2	98.0	2.0	4	2.0 " "
97.0	3.0	65	1.0	96.0	4.0	8	0.8 " "
95.0	5.0	90	5.9	94.0	6.0	13	1.6 " "
93.0	7.0	120	7.9	91.5	8.5	0 (eutectic)	2.0 " "
90.0	10.0	150	7.5-10.0	90.0	10.0	4	2.7 " "
85.0	15.0	160	7.9-8.7	87.5	12.5	3	0.6 " "
80.0	20.0	100	9.4-11.0	—	—		

atm. If the specimen shows no leakage, 1 or 2 mm was machined off both its ends (off each end alternately) and the imperviousness test repeated. The value of imperviousness and consequently amount of scattered porosity was determined as the residual "critical" thickness of the middle part (wall) of the casting at which "seepage" appeared at the given pressure.

For ascertaining the relationship between composition and imperviousness of alloys, three series of alloys of the binary systems Cu—Sn, Cu—Al and Cu—Si were tested. The test specimens were cast in sand molds at a temperature exceeding the liquidus by 75 - 100°C. The dimensions of the specimens were 135 × 60 × 25 mm, and the lower half of the casting (70 × 60 × 25 mm) free from piping was subjected to test.

It has already been shown, in considering the dependence of hot-shortness of alloys on their composition, that alloys solidifying in a considerable temperature range are in the liquid-solid state for a long time, much longer than the time for complete solidification of the pure component of the alloy or the eutectic (see Figs. 75 and 76). The crystallization of such alloys, for different conditions of heat transmission from the solidifying mass, results in the formation of more or less equiaxial crystal dendrites having a relatively large internal interface and, consequently, susceptible to the formation of enclosed regions, difficult of access for the residual liquid feeding and the casting.

In these circumstances, with increase in the solidification range of of the alloy and in the width of the zone of the casting affected by shrinkage porosity, the "critical" thickness of hydraulically tested castings is also increased. Table 33 gives the results of hydraulic resistance tests on alloys of copper with tin and aluminum.

From a comparison of the hydraulic resistance values for these two series of alloys, it may be concluded that the volume of scattered shrinkage pores is directly related to the composition and crystallization range. Copper-aluminum alloys, crystallizing in a very narrow temperature range, failed to show characteristic leakage at any composition; at the final wall thickness of about 1 mm and a pressure of 60 - 80 atm, the wall bulged or burst. Isolated, well-visible pores (diameter 0.5 - 0.1 mm) of a shrinkage or gas character, seen in aluminum-bronze castings during the gradual machining of the latter, did not cause leakage. These defects are evidently of a strictly local character and are not connected with interdendritic microscopic porosity, the presence of which is inherent in alloys crystallizing over a considerable temperature range. In specimens of copper alloys with 5% tin and more, leakage is found at a wall thickness of 6 - 8 mm; the water first appears in the form of "seepage" over the entire face of the specimen, and the alloys as it were "swell." Not infrequently, after a certain time, during which the pressure is maintained constant at 30 - 40 atm (or as the pressure increases), cracks are formed from which the water issues in jets.

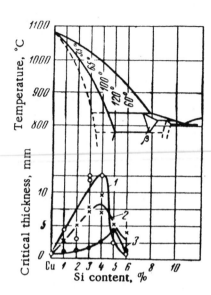

Fig. 129. Variation in hydraulic resistance of Cu—Si alloys as function of composition and method of casting the specimens: 1) Without presolidification; 2) with presolidification; 3) chill cast.

Cu—Si alloys were found to have the lowest hydraulic resistance. The introduction of a small quantity of silicon (0.75 - 1.0%) into pure copper (which if properly cast is perfectly resistant to hydraulic pressure), results in leakage at a residual wall thickness of 3 - 4 mm. An increase in silicon content to 4% leads to a further reduction in hydraulic resistance, and only at 6% Si is it possible to produce comparatively tight castings. Figure 129 shows curves for three different methods of casting these alloys; all three curves have a maximum, corresponding to maximum volume of scattered shrinkage porosity. The ordinary method of casting the alloys immediately after melting the charge (without presolidification) resulted in a low hydraulic resistance of the castings. During test, local spouting leaks (cracks) were observed, which were sometimes due to the presence of oxide films and large gas blowholes. In the second series of experiments, therefore, "presolidification" was used, that is to say the liquid alloys were cooled to solidification and this was followed by rapid re-melting. The results obtained with this method showed that alloys with 3 - 4% Si, crystallizing in a considerable temperature range, are more affected by porosity than the other alloys of this series. The third series of alloys (chill cast) did not affect these results, except that in this case, the hydraulic resistance maximum was much lower and was shifted toward the equilibrium maximum of the crystallization range.

In comparing the results in hydraulic resistance tests of Cu—Sn and Cu—Si alloys with the form of the constitutional diagrams of these alloys, it should be pointed out that maximum hydraulic perviousness of the alloys, i.e., maximum value of scattered shrinkage porosity, practically corresponds to the saturation limit of α-solid solution, which in its turn coincides with maximum crystallization range. The difference in behavior of these two series of alloys is that in Cu—Sn alloys, no increase in hydraulic resistance is observed with increase in the crystallization range (with increase in the tin content of the alloys), while in the case of Cu—Si alloys, this increase in hydraulic resistance is shown definitely by the downward slope of the curve in the region of 5 -6% Si, corresponding to a decrease in the temperature range of solidification, and has a favorable effect in reducing porosity.

In subsequent investigations, it was found also that alloys of the Cu—Sb system, crystallizing in a considerable temperature range (for some alloys as much as 350 - 390°C), were found to have perfect hydraulic resistance, showing the absence of shrinkage porosity in the cast specimens. It is evident that this is due to the good impregnation qualities of Cu—Sb alloys compared with Cu—Sn and Cu—Si alloys, which may be helped by the low surface tension. There are no definite figures for the value of this property, but it has been observed that antimony considerably reduces the surface tension of copper. At the same time, probably, the increase in wettability of the crystals by the liquid melt facilitates the penetration of residual liquid into the shrinkage voids. If this is so, it may be said that there is a connection between the surface properties of alloys in the liquid and liquid-solid states and microscopic porosity in castings. This, however, would require further proof. Nevertheless, the fundamental relationship which can be observed in the distribution of shrinkage voids in alloys in association with their position in the constitutional diagrams is fully established on the basis of earlier work on aluminum alloys [85, 86, 226], and on the basis of this work on copper alloys.

Definite confirmation of the above-mentioned character of the variation of shrinkage voids in castings of different alloys is to be found in an analysis of the fractures and microstructures of specimens tested for hydraulic resistance. Such an analysis shows that castings of copper and aluminum bronze, which have crystallized under identical thermal conditions, have a structure consisting of columnar dendrites, growing towards the center of the casting, while in the case of castings of silica bronze and tin bronze, which have solidified in a temperature range, the zone of columnar crystals occupies a relatively small volume in the outer part of the casting, while its central part consists of more or less equiaxial dendrites. Thus, the cause of the low hydraulic resistance of Cu–Sn and Cu–Si alloys of a certain composition is shrinkage porosity, produced at the numerous joints of the branching dendrites and resulting from deficient feeding; the characteristic structure pertaining to these two cases has already been illustrated (see Figs. 126 and 127).

The results quoted for the porosity of copper alloys refute the point of view previously expressed by some investigators to the effect that alloys intended for castings should have a considerable temperature range of crystallization [228].

Dependence of Shrinkage Porosity ("Black Fracture") on Composition in Magnesium Alloys

Subsequent work on the study of the distribution of internal shrinkage voids was carried out on alloys of the Mg–Al system. The need for such work arose from the situation that industrial magnesium alloys are the lightest of all the alloys used in machine construction, and the prospects of their increase use grow annually. This is also assisted by work aiming at the continued improvement of the corrosion resistance of magnesium alloys.

One of the principal defects in castings of magnesium alloys is the presence in the fracture of castings, especially in transitional parts, of regions of microporosity, considerably reducing the physical and mechanical properties of

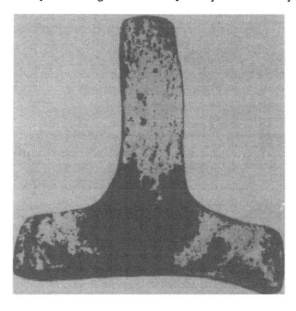

Fig. 130. Characteristic appearance of "black fracture" in a magnesium alloy casting.

the castings [220, 230 and 231]. This defect is met with in castings produced from the most widely used magnesium casting alloy ML-5, containing 8 - 9% Al. Figure 130 shows the characteristic appearance of a casting with "black fracture." The mechanical properties of specimens cut from this portion of the casting were 30 - 40% below those of specimens from adjacent parts showing a normal fracture.

The constitutional diagram of the Mg–Al system (Fig. 131) shows that alloys of the above-mentioned composition solidify in the maximum temperature range for this system, amounting to 165°C in the case of nonequilibrium crystallization. This produces a loose shrinkage structure and porosity. Nevertheless, alloys having such an aluminum content are widely used, since they have high strength coupled with good ductility. Although information on the dependence of the strength of Mg–Al alloys on their composition was already known, we carried out some fresh work to ascertain the composition of alloys showing the best combination of strength and ductility.

Tests were carried out on two series of specimens in the as cast condition (sand mold) and after heat treatment, consisting of homogenizing at 410 ± 5°C for 14 hr followed by air cooling. The results of the tests are shown in Fig. 132. The curves for ultimate strength and elongation show that the best combination of these properties is to be found for a content of about 8% Al in the alloy. The improving effect of homogenizing is due to the solution of the brittle structural component consisting of the chemical compound Mg_4Al_3, and to the supersaturation of the solid solution by aluminum.

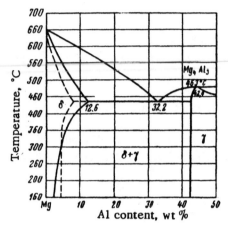

Fig. 131. Constitutional diagram of Mg–Al system.

Microporosity ("black fracture") in castings of this alloy is, of course, extremely undesirable, since it reduc the strength and other properties. Efforts to prevent this defect are complicated by the fact that so far there has be no definitely decided opinion as to whether "black fracture" was due to shrinkage porosity or to liberation of gas. At first sight it would appear that this is immaterial, since both shrinkage and gas liberation lead to the same resul but it is essential to ascertain the main cause in order to solve the problem as to methods of preventing this defect.

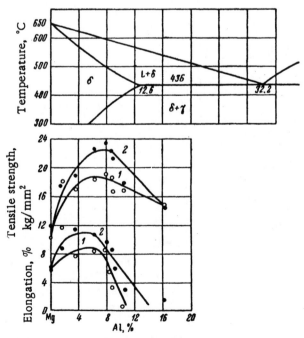

Fig. 132. Dependence of mechanical properties of Mg–Al alloys on composition: 1) As cast; 2) after heat treatment.

Thus, if "black fracture" is due to the liberation of gas, it may be eliminated by vacuum treatment of the liquid metal before it is poured into the mold or by some other degasifying treatment; if, however, the principal cause is shrinkage porosity, the best method may be to use the method of crystallization under pressure, vibrational agitation, etc. Since usually both causes are operative simultaneously, the method recommended by A. A. Bochvar and A. G. Spasskii of vacuum suction of the metal into the mold followed by disconnection of the vacuum and the application of pressure during solidification of the casting may be used for producing sound castings.

In our work, the specimens for the examination of the fractures for external appearance and microstructure were cast from alloys (Table 34) prepared in an electric furnace of semi-industrial type. Melting of the metal, casting of the specimens and heat treatment were carried out in accordance with the accepted technical production of castings in industrial conditions.

TABLE 34. Composition of Magnesium Alloys According to Charge and According to Analysis

Composition according to charge, %	Composition according to analysis, %			Composition according to charge, %	Composition according to analysis, %		
Al	Al	Si	Fe	Al	Al	Si	Fe
Mg, commercial	0.06	0.02	0.02	8.0	8.05	0.05	0.09
2.0	3.07	0.06	0.07	9.0	8.82	0.02	0.02
2.2	2.75	0.01	0.03	9.5	9.96	0.05	0.1
3.0	3.71	0.02	0.015	11.0	11.01	0.13	0.063
4.0	5.22	0.05	0.03	11.0	10.26	—	—
6.0	6.43	0.03	0.04	15.0	14.84	0.12	0.054
6.0	6.20	0.02	0.03	15.0	16.32	0.03	0.02
				20.0	19.74	0.11	0.054
				30.0	29.48	0.19	0.063

Two blocks (Fig. 133a), three corrugated bars (Fig. 133b), and four brackets (Fig. 133c) were cast from each melt. These items were selected for the experiments because the position of the regions of microporosity in them was known from previous examinations of fractures. By comparing the appearance and size of the portions of the castings affected by microporosity, it was possible to draw conclusions regarding the systematic influence of alloy composition on the quantity of shrinkage voids in the test specimens, since they were cast at temperatures equidistant from the liquidus temperature (100 - 120°C above the liquidus), and the solidification conditions of each of the three bars and four brackets in one mold could be regarded as identical. The final castings were always fractured at definite cross sections, as shown in Fig. 133. One of the castings was tested in the as cast condition and the other two after heat treatment. The blocks (Fig. 133a), which were poured without special gates (but with intensified feeding) were examined for fracture in the heat-treated condition.

Figure 134 shows the nature of the fracture of the blocks. In this case, where the casting was of very simple shape and the method of feeding was also very simple, visible microporosity was found for an Al content in the alloy of 8.05%. Fractures of castings of pure magnesium and low aluminum content (up to 6% Al) were free from porosity and its accumulations in transitional sections. No microporosity was to be found also in castings of an alloy containing 29.48% Al. Thus, in the Mg–Al system, there are alloys in which the susceptibility to the formation of scattered and concentrated porosity is high, and alloys in which that susceptibility is low, as was also found to be the case with aluminum and copper alloys.

In the examination of the structure of fractures and the quantitative assessment of regional porosity in sections of the castings, it was found expedient to rate porosity on a four-point basis (Fig. 135). If the cross-sectional area was affected by porosity by up to 10%, this was denoted by a porosity rating of 1, if the area was affected within the limits of from 10 to 25%, by a porosity rating of 2, in the limits from 25 to 40%, by a porosity rating of 3 and if 40% or more of the cross-sectional area was affected, this was denoted by a rating of 4.

The point assessment of the susceptibility of an alloy of given composition to produce castings with microporosity was made for the whole of the bar casting additively for all eight sections. Thus, of the eight sections examined in a casting, if three had a microporosity rating of 3 and two had ratings of 1 and 2, respectively, the total porosity of the entire casting was rated at 12. Susceptibility to microporosity in the examination of the bracket casting was rated in the same way (according to the three sections shown in Fig. 133c).

The results of careful examinations of the fractures showed that in bar castings of pure magnesium and alloys containing 2.75 and 3.7% Al, no microporosity was to be found in any section. Maximum microporosity was found in castings of alloys with 8.82 and 10.26% Al. It should be noted that the most definite results in the rating of microporosity are obtained in the analysis of fractures of castings which have been heat treated, i.e., subjected to heating for a long time in closed boxes. In castings which have not undergone such heating, microporosity is revealed with difficulty; only in the case of strongly developed porosity in the fracture is it possible to see a light-gray or gray coloring in the central region of the fracture. If, however, the castings are heated in closed boxes in an air furnace, the sections of the castings in which microporosity was produced during solidification become dark in color. During heating, the porosity region or zone is as it were "developed" by the reaction of the large surface of the unsound metal with the oxygen and nitrogen of the air, resulting in the formation of magnesium oxides and nitrides of different thicknesses and consequently different degrees of coloring. The intensity of the coloring increases with increase in the porosity-affected area of the casting.

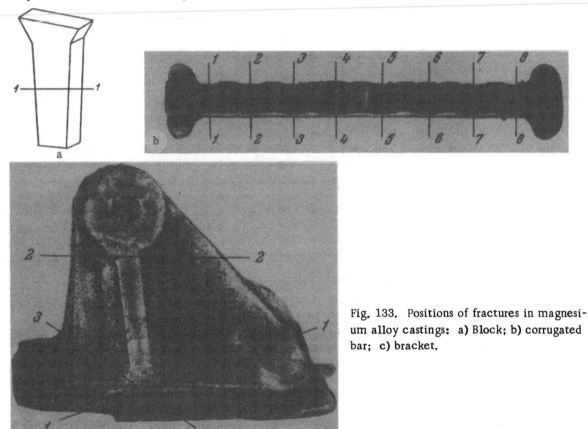

Fig. 133. Positions of fractures in magnesium alloy castings: a) Block; b) corrugated bar; c) bracket.

The results of fracture assessment reproduced in Fig. 136 show that the total porosity rating (i.e., the proportion of the area of the fracture affected by shrinkage porosity) increases with increase in the aluminum content of the alloy up to 11 - 15%, and then diminishes. Microporosity is completely absent in alloys solidifying at constant temperature, for example in castings of pure magnesium and the alloy with 29.48% Al, i.e., an alloy near the eutectic in composition (32% Al). Alloys with a low aluminum content either have no porosity at all or are only slightly affected by it. Maximum shrinkage porosity is found in alloys solidifying in the maximum temperature range (for the case of the nonequilibrium constitutional diagram).

Thus, the results of the investigations once more confirm the existence of quite a definite relationship between the composition of alloys, i.e., their position in the constitutional diagram, and the character and quantity of shrinkage voids in castings. In alloys solidifying at constant temperature, a shrinkage pipe or cavity without scattered porosity is formed; in alloys having a considerable solidification range, shrinkage appears in the form of scattered or regional porosity. The mechanism of its production is directly related to the character of the solidification of magnesium solid solution of variable composition, when the intensely branching dendrites do not receive an adequate supply ot residual liquid melt to compensate the shrinkage. As the aluminum content of the alloy increases, the relative quantity of liquid towards the end of solidification also increases, the "impregnation" of the casting is im-

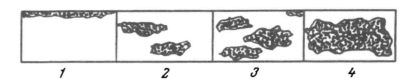

Fig. 135. Rating of "black fracture" on the four-point system.

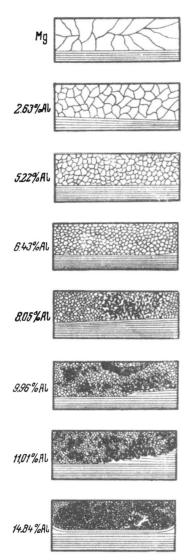

Fig. 134. Appearance of "black fracture" in castings of Mg–Al alloys with variation in composition.

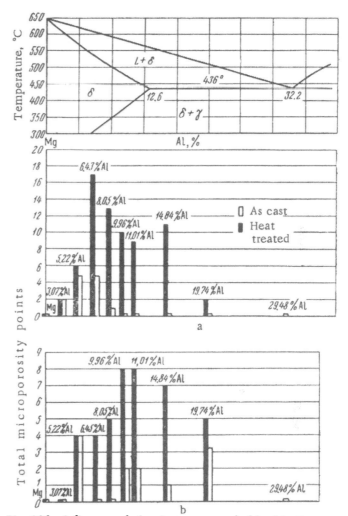

Fig. 136. Influence of aluminum content in Mg–Al alloys on the formation of microporosity in castings: a) Total microporosity rating in the eight sections of a bar casting; b) the same in three sections of a bracket casting.

proved, i.e., the penetration of liquid between the adjacent growing dendrites and also between the branches of the dendrites themselves. The increased supply of liquid metal counteracts the natural susceptibility of such alloys to microporosity. This was found to be the case in experiments with a bracket casting having an increased feeder head, i.e., with increased feed of a casting in which there is usually concentrated microporosity. It is thus possible to obtain sound, nonporous castings with 5 - 7% aluminum, while is feeding is inadequate, castings of alloys of the same composition have appreciable porosity.

Microscopic examination of polished specimens from regions of concentrated porosity confirms the mechanism of the formation of shrinkage porosity as described above. The porous areas under the microscope show structural imperfections either in the form of rough, oxidized grain boundaries (Fig. 137) or in the form of the typical porosity (Figs. 138 and 139). It will be seen that these areas are deprived of eutectic component present in insufficient quantity to feed these portions. Usually, the size of the individual microporosity areas does not exceed 0.8-1 mm in cross section, but the accumulation of large numbers of such areas forms "black" zones or regions of microporosity of considerable extent, well visible to the unaided eye.

The examination of the microstructure of numerous specimens of magnesium-aluminum alloys of different compositions shows that the commonest form of microporosity is typical of shrinkage but not of the liberation of gas. It is not always possible, however, to distinguish between shrinkage porosity and gas porosity [232, 233], this being quite understandable, since both phenomena, shrinkage and gas liberation, occur together during solidification, and

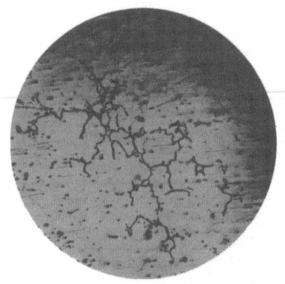 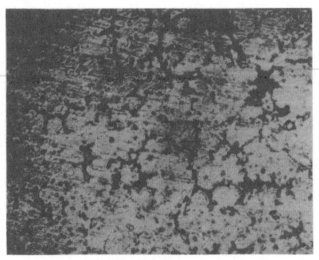

Fig. 137. Area of microstructure with oxidized grain boundaries.

Fig. 138. Shrinkage microporosity in Mg-Al alloys.

it is very probable that the liberated gas tends to fill the voids produced in the form of shrinkage pores. If the amount of gas (under high pressure) is considerable, the shrinkage pores may be deformed until they are spherical; in places where there is an appreciable quantity of residual liquid, the latter may displace the gas and fill some of the shrinkage cavities, while the rest of the cavities remain filled with gas. It is obvious that new, suitable methods will be required for the full investigation of castings in regard to the origin and classification of such internal defects.

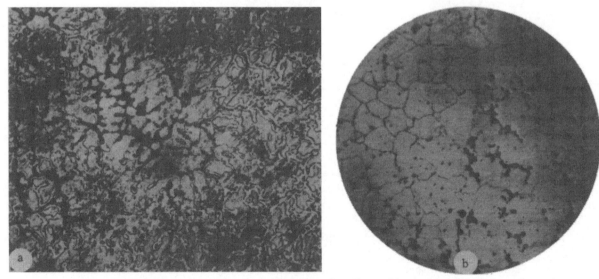

Fig. 139. Shrinkage microporosity in the alloy ML-5: a) Along the dendrite boundaries; b) point microporosity.

To summarize the foregoing, it may definitely be concluded that in alloys solidifying at constant temperature, practically the entire volume of the internal shrinkage appears in the form of a concentrated shrinkage cavity or pipe, which, if the mold is properly designed and the solidification conditions of the casting are correctly selected, may be located completely outside the casting. In alloys solidifying in a considerable temperature range with the formation of branching dendrites of primary crystals, and with a relatively small amount of residual liquid phase, shrinkage appears in the form of regions of porosity, porosity scattered throughout the entire body of the casting, and cracks. The results obtained show that the relative quantity of liquid towards the end of crystallization adequate for reducing and eliminating shrinkage microporosity in castings ought to be 15 - 25%; this figure is close to that found by A. A. Bochvar for some aluminum casting alloys [208]. This is confirmed by the examination of "composition – hydraulic resistance" diagrams for copper-based and aluminum-based alloys, "composition – microporosity" curves for magnesium-based alloys, and also recent work on the crystallization of Al–Mg alloys in autoclaves under a pressure of 4 - 5 atm [234].

These conclusions follow from the form of solidification of alloys occupying a definite position in the constitutional diagrams the physical nature of which is unique for metals. External factors, such as temperature of the metal and mold, rate of pouring and solidification, degasification of alloys before or during pouring, may modify the "natural" properties of alloys. In isolated cases, by varying the external factors, the negative properties of alloys can be entirely suppressed and castings of perfectly good quality obtained. The search for such possibilities is a problem for research workers and foundry technicians, and its solution will result in the continued improvement in casting techniques.

CONCLUSION

The extensive development of methods and processes for the industrial production of machine parts direct from the liquid metal and the expansion of ingot production by continuous casting methods call for a reliable knowledge of alloy theory. Such a section of the metallurgy of foundry practice may be established in direct connection with the experimental study of "composition –casting property" diagrams, in the same way in which theories have been and are being established for various alloy properties, such as theories of strength, plasticity and the like.

The considerable amount of data relative to the casting properties of alloys and appearing on an ever-increasing scale in the published literature requires generalization on a physicochemical basis, i.e., in relation to the constitutional diagrams of alloys.

It is on the basis of the relationships expressed in the form of "composition – casting property" diagrams and their study in relation to the constitutional diagrams that the most rational composition of casting alloys with the best technical and service properties can be selected, and the search for new alloys can be conducted with minimum expenditure of time and means.

The complex formed by the casting properties of alloys, set out in the form of "composition – property" diagrams, expresses the character of alloys in the liquid, liquid-solid, and solid states, and enables the data acquired in industrial practice to be generalized.

Properties of Metals and Alloys in the Liquid State

The surface tension of pure metals is characterized by the different values of the energy of the atoms and electrons in the body of the liquid and on its surface. The surface tension of metals can be calculated by formulas taking this energy into account and derived from the theory of the metal state and the quasicrystalline structure of liquids. The best agreement between calculated and experimental values of the surface tension of metals is obtained when it is calculated from formulas which take into account the density and specific gravity of the metals or the electron work function and the atomic radius.

The more exact values which have been obtained for the surface tension of aluminum, zinc and other metals show that the surface tension of metals is in direct relation to their atomic volume.

The relationships in the variation of surface tension as a function of composition and the form of the constitutional diagrams which we have studied for a number of aluminum-based and zinc-based binary alloys enable the following conclusions to be drawn:

1. Alloying of the solvent metal (aluminum or zinc) with various other metals results in a reduction in surface tension of the solvent in those cases where the added metal has an atomic volume greater than the atomic volume of the solvent metal (Al–Li, Al–Ca, Al–Mg, Al–Pb, Zn–Sb, Zn–Sn, Zn–Bi, Mg–Sb, Mg–Sn and other systems, see Figs. 29 - 32).

2. The surface tension is considerably reduced in the case of comparatively low concentrations of a surface-active metal (addition).

3. If the added component has an atomic volume which is equal to or less than that of the solvent, the surface tension of the alloys is not appreciably affected (Al–Si, Al–Cu, Al–Zn, Al–Fe, Zn–Cu, Zn–Ag, Mg–Al, Mg–Mn and other systems).

4. The existence of a chemical reaction between the alloy components, resulting in the formation of chemical compounds on solidification, is not always accompanied by a variation in surface tension (Al–Cu, Al–Fe and other alloys).

5. The variation in surface tension of ternary alloys is similar to its variation in binary alloys; the reduction in surface tension on the addition of surface-active components is proportional to the difference in the atomic volumes of solvent and dissolved metal and to the concentration of the latter (Al–Cu–Mg, Al–Si–Mg, Al–Zn–Mg, Al–Si–Pb and other systems, see Figs. 35, 37, 38). The existence of a chemical reaction between the components and additions increases the effect of reducing the surface tension (Al–Mg–Pb, Al–Mg–Bi systems, see Fig. 39).

Calculations of the adsorption of surface-active components on aluminum and zinc give characteristic curves of adsorption as a function of concentration, with a maximum which is more pronounced, the greater the activity of the added component (Al–Li, Al–Ca, Zn–Sb, Zn–Sn and other systems, see Figs. 40 and 41). Calculation of the adsorption may be a method of approximately estimating the sizes of the particles in the adsorbed layer.

The viscosity of pure metals, like their surface tension, depends on the energy of interaction of the atoms and may be calculated from formulas which take this energy into account. A comparative estimate of the kinematic viscosity of metals may be made on the basis of the values for the atomic volume or the standard entropy of metals; the higher these values are, the lower is the viscosity.

For many metals, the value of the kinematic viscosity may be calculated from the empirical formula proposed by the author (see p. 62), which takes the atomic volume of the metal into consideration.

A systematic measurement of the viscosity of eutectic alloys indicates the existence of minimum viscosity in alloys of eutectic composition or close to it (Al–Si, Al–Cu and other alloys).

Casting Properties of Alloys in the Liquid-Solid and Solid States

The fluidity of pure metals is greater, the greater their heat of crystallization. As a rule, impurities reduce the fluidity of metals (see Figs. 65 and 66).

"Composition – fluidity" diagrams plotted for a number of binary systems definitely show that maximum fluidity of alloys corresponds to a eutectic composition and chemical compounds in the constitutional diagrams, i.e., to alloys solidifying at constant temperature; minimum fluidity is found in alloys solidifying in a considerable temperature range (see Figs. 69 - 74). This is due to the quantity of heat of primary and secondary crystallization of alloys of different compositions, and also to a structural factor. A needle-like form of primary crystals considerably reduces the fluidity of alloys.

The causes of the regular variation in fluidity of alloys of ternary systems are the same as for alloys of binary systems, i.e., thermal and structural factors. Maximum fluidity is found in alloys corresponding to the composition of the ternary eutectic point (Al–Cu–Si and Al–Mg–Zn alloys, see Figs. 83 and 84).

Despite the widely held view, fluidity does not depend on surface tension or the kinematic viscosity of metals and alloys in a definitely liquid state.

Investigations of shrinkage phenomena have shown the following results of the greatest interest for alloy theory and foundry practice.

The time for complete solidification of alloys crystallizing in a temperature range, other conditions being equal, is longer than for the pure components or alloys solidifying at constant temperature or in a narrow temperature range. This is due both to the reduction in the solidus temperature and to the relatively low thermal conductivity of alloys in the liquid and liquid-solid states (see Figs. 75 and 76).

However high may be the rate of solidification of the surface and deep layers of castings solidifying in a considerable temperature range and in the presence of an effective range, they pass through a phase of solid-liquid state, accompanied by linear shrinkage. If shrinkage is hindered, crystallization cracks are produced. Pure metals and eutectics are not susceptible to crystallization cracking.

The systematic variations found for shrinkage in alloys of many binary systems not subject to transformations in the solid state make it possible to calculate linear shrinkage from the coefficients of thermal expansion and the solidus temperature of the alloys.

From the character of the shrinkage curves of alloys when compared with the constitutional diagrams, it is possible to indicate a region of alloy temperatures and concentrations, within the limits of which exist solid-liquid mixtures having high susceptibility to hot-shortness on solidification.

The "composition – hot-shortness" curves constructed by available methods for alloys of many binary systems point to a definite relationship in the variation of this property as a function of the composition and form of the constitutional diagrams. Maximum hot-shortness is found in alloys which solidify in the maximum "effective" temperature range and retain about 10% of residual liquid towards the end of crystallization (see Figs. 109 - 111).

With a certain variation in the composition of hot-short alloys, the crystallization cracks produced by hindered shrinkage are found to be healed by being filled with residual liquid, the relative quantity of which towards the end of solidification, for a number of alloys, ought not to be less than 20%.

The relationships in the variation of shrinkage and susceptibility to hot-shortness observed in alloys of binary systems are also inherent in alloys of ternary systems (see Figs. 117 - 120).

On the basis of research results, it is possible to indicate measures for counteracting hot cracks in castings of various alloys. One such measure is a variation in the alloy composition, sometimes not exceeding the limits imposed by standards.

The regular distribution of other shrinkage defects, such as shrinkage cavities and scattered porosity, also depends on the composition of the alloys and their position in the constitutional diagrams.

In alloys solidifying at constant temperature, the entire volume of internal shrinkage appears in the form of a concentrated shrinkage cavity or pipe. Scattered shrinkage porosity is found in alloys solidifying in a considerable temperature range, due to the character of the crystallization and structure of such alloys (see Figs. 126, 127, 138, 139).

LITERATURE CITED

1. V. V. Danilevskii. Russian Technology [in Russian] (Lenizdat, 1957).
2. N. N. Rubtsov. "From Andrei Cokhov to Ivan Motorin (Russian bell-founding and gun-casting art of the end of the 16th and the commencement of the 18th centuries," Collection: Technology of Foundry Practice [in Russian] (Mashgiz, 1955).
3. I. N. Bogachev. P. P. Anosov and the Secret of Damask Steel [in Russian] (Mashgiz, 1952).
4. Russian Metallurgical Scientists. Collection of Articles [in Russian] (Mashgiz, 1951).
5. P. P. Anosov. Collected Works [in Russian] (Izd. AN SSSR, 1954).
6. A. Ya. Chernyak and D. M. Nakhimov. The Russian Metallurgical Scientist N. V. Kalutskii [in Russian] (Mashgiz, 1951).
7. A. G. Spasskii. Principles of Foundry Practice [in Russian] (Metallurgizdat, 1950).
8. N. G. Girshovish. Iron Castings [in Russian] (Metallurgizdat, 1949).
9. A. S. Lavrov. Cited from [14].
10. D. K. Chernov and the Science of Metals [in Russian] (Metallurgizdat, 1950).
11. N. T. Gudtsov. "Fundamental problems in the study of the steel ingot," [in Russian] (1952).
12. A. A. Bochvar. "Production and casting of alloys," Collection: Nonferrous Metallurgy of Central and East Germany [in Russian] (Metallurgizdat, 1947).
13. V. I. Dobatkin and E. T. Safonova. "Production of circular ingots of more than 500 mm diameter," Collection: Aluminum Alloys [in Russian] (Oborongiz, 1955).
14. N. E. Osokin. "Casting of ingots by the method of A. Lavrov," Collection: Technology of Nonferrous Metals [in Russian] (Metallurgizdat, 1952).
15. E. Hein. Metallography in Its Application to Metallurgy [Russian Translation] (1904).
16. A. A. Baikov. Collected Works, Vol. 2 [in Russian] (Izd. AN SSSR, 1948).
17. A. A. Bochvar and A. G. Spasskii. Aviapromyshlennost' No. 7 (1936).
18. V. M. Plyatskii. Casting Processes Using High Pressures [in Russian] (Mashgiz, 1954).
19. L. I. Fantalov and L. I. Levi. "Application of closed feed heads in the production of castings," Collected Works: Technology of the Production of Steel and Alloys, Trans. Moscow Steel Institute [in Russian] (Metallurgizdat, 1946); L. I. Fantalov and L. I. Levi. Application of Closed Shrinkage Heads Acting under Atmospheric Pressure in the Production of Castings [in Russian] (Inst. tekhniko-ékonomicheskoi informatsii Gosplana SSSR, No. 1, 1946).
20. I. A. Oding et al. "Investigation of aluminum and nickel-aluminum bronzes," Collection: Metallurgical Research [in Russian] (Gosmatmetizdat, 1932).
21. V. E. Neimark. "Production of castings by the method of vacuum crystallization," Collection: Solidification of Metals [in Russian] (Mashgiz, 1958).
22. I. I. Novikov et al. "Application of vibration in crystallization for eliminating hot cracks," Liteinoe proizvodstvo No. 1 (1958).
23. R. S. Richards and W. Rostoker. "The influence of vibration on the solidification of an aluminum alloy," Trans. Amer. Soc. Metals 48 (23), 884 (1956).
24. N. T. Gudtsov. "On the question of improving the structure of a steel ingot," Tr. NTO chernoi metallurgii, Vol. 5, part II (1955).
25. A. M. Samarin and L. M. Novik. Use of Vacuum in Steel-Melting Processes [in Russian] (Metallurgizdat, 1957).
26. A. A. Bochvar. Casting Properties of Alloys. Collection of Scientific Papers [in Russian] (Izd. BNITOM, 1940).
27. N. S. Kurnakov. Introduction to Physicochemical Analysis [in Russian] (Izd. AN SSSR, 1940).
28. Ya. I. Frenkel'. Kinetic Theory of Liquids [in Russian] (Izd. AN SSSR, 1945).
29. V. I. Danilov. Scattering of X-Rays in Liquids [in Russian] (ONTI, 1935).
30. E. Bartholome. "Structure and intramolecular forces in pure liquids and solutions," Report of a Meeting of the Faraday Society in 1935. Usp. khimii 6, No. 6 (1937).

31. P. Debye. "Methods of determining the electrical and geometrical structure of molecules," Usp. khimii 6, No. 5 (1937).

32. V. K. Semenchenko. Physical Theory of Solutions [in Russian] (GITTL, 1941).

33. V. K. Semenchenko and N. L. Pokrovskii. "Surface tension of molten metals and alloys," Usp. khimii 6, 6-7 (1937).

34. V. I. Danilov and I. V. Radchenko. "Structure of liquid metals near the crystallization point," ZhETF 7, 1152 (1937).

35. V. I. Danilov and I. V. Radchenko. "Scattering of x-rays in liquid eutectic alloys," ZhETF 7, 1158 (1937).

36. V. I. Danilov and A. I. Danilova. "X-ray study of liquid alloys. Method. Bismuth-lead alloy," Collection: Problems of Metallurgy and Physics of Metals, No. 2 [in Russian] (Metallurgizdat, 1951).

37. K. P. Bunin. "On the question of the structure of metallic eutectic alloys," Izv. AN SSSR, OTN No. 2 (1946).

38. K. P. Bunin and Ya. N. Malinochka. Introduction to Metallography [in Russian] (GNTI, 1954).

39. A. A. Bochvar. Mechanism and Kinetics of the Crystallization of Alloys of Eutectic Type [in Russian] (ONTI, 1935).

40. A. A. Bochvar. Science of Metals [in Russian] (Metallurgizdat 1956).

41. E. G. Shvidkovskii. Some Problems of the Viscosity of Molten Metals [in Russian] (GITTL, 1955).

42. Ya. S. Umanskii et al. Physical Principles of the Science of Metals [in Russian] (Metallurgizdat, 1955).

43. B. R. T. Frost. Structure of Liquid Metals [Russian translation] Uspekhi fiziki metallov (Metallurgizdat, 1958).

44. V. I. Danilov and B. E. Neimark. "The presence of crystallization nuclei above the melting point and the structure of liquids, ZhETF 7, 1161 (1937).

45. P. B. Dankov. "Mechanism of phase conversions from the point of view of orientational dimensional correspondence," Izv. SFKhA 16, No. 1 (1953).

46. P. D. Dankov. "Crystal-chemical mechanism of the reaction of a crystal surface with foreign elementary particles," ZhFKh 20, 853 (1946).

47. V. I. Danilov and D. E. Ovsienko. "Nucleation of centers on active impurities," ZhETF 21, 879 (1951).

48. V. I. Danilov and A. G. Pomogaibo. "On the crystallization of sodium and potassium," Dokl. An SSSR 68, No. 5, 843 (1949).

49. V. I. Danilov. "Some problems of the kinetics of the crystallization of liquids, Collection: Problemy Metallovedeniya i fiziki metallov [in Russian] No. 1, 7 (1949).

50. A. V. Shubnikov. How Crystals Grow (Izd. AN SSSR, 1935).

51. I. N. Fridlyander. Study of the Form of Growth of Crystals as a Function of the Rate of Cooling [in Russian] (Oborongiz, 1949).

52. "Replies to a questionnaire on dendrites," Metallurg Nos. 7 and 8 (1935).

53. B. B. Gulyaev. Solidification and Inhomogeneity of Steel [in Russian] (Metallurgizdat, 1950).

54. D. D. Saratovkin. Dendritic Crystallization [in Russian] (Metallurgizdat, 1953).

55. B. B. Gulyaev. "Solidification and inhomogeneity of a killed steel ingot," Collection: The Steel Ingot (Metallurgizdat, 1952).

56. V. I. Dobatkin. Continuous Casting and the Casting Properties of Alloys [in Russian] (Oborongiz, 1948).

57. I. N. Fridlyander. "Study of the influence of cooling rate on the structure and properties of aluminum alloys," Collection: Solidification of Metals (Mashgiz, 1958).

58. Collection: Growth of Crystals (Consultants Bureau, N. Y., Vol. I, 1958; Vol. II, 1959).

59. N. A. Bushe. "Influence of chemical composition of calcium babbitt metal on the service properties of aluminum alloys," Collection: Research on Nonferrous Metal Alloys [in Russian] (Izd. AN SSSR, 1955).

60. D. A. Petrov et al. "Production of germanium and silicon single crystals," Collection: Problems of the Metallurgy and Physics of Semiconductors [in Russian] (Izd. AN SSSR, 1957).

61. G. L. Livshits. "On the question of the formation of the bottom cone in an ingot," Stal' No. 6, 518 (1952).

62. I. N. Golikov and F. V. Kozlov. "On the question of shower crystallization in steel," Stal' No. 7, 626 (1952).

63. G. P. Ivantsov. "On the question of the formation of showers of crystals in a steel ingot," Stal' No. 10, 922 (1952).

64. V. M. Tageev. "Hypothesis concerning showers of crystals in solidified ingots and castings," Stal' No. 1, 59 (1952).

65. G. N. Oiks. "Questions concerning the crystallization of a steel ingot," Stal' No. 8, 735 (1952).

66. N. E. Skorokhodov. "On the question of showers of crystals," Stal' No. 9, 824 (1952).

67. I. N. Fridlyander. "Investigation of bright crystals in continuously cast aluminum alloy ingots," Liteinoe proizvodstvo No. 10 (1956).

68. V. I. Dobatkin. "Structure of ingots of deformed aluminum alloys and its influence on the properties of the products," Dissertation for Doctor's Degree [in Russian] (Mintsvetmetzoloto, 1956).

69. V. I. Danilov and G. Kh. Chedzhemov. "Influence of supersonic vibrations on the crystallization of super-cooled liquids and the formation of primary crystallization structure," Collection: Problemy Metallovedeniya i fiziki Metallov [in Russian] No. 4 (1955).

70. V. Ya. Anosov and S. A. Pogodin. Fundamental Principles of Physicochemical Analysis [in Russian] (Izd. AN SSSR, 1948).

71. A. A. Bochvar and K. V. Gorev. a) "Abnormal structures in slowly cooled alloys of eutectic type," b) "Crystallization of ternary eutectics," Collected Works of the Metallographical Laboratory of Mintsvet metzolota. [In Russian] (ONTI, Metallurgizdat, 1933); A. A. Bochvar and O. S. Zhadaeva. "Microstructure of hypoeutectic and hypereutectic alloys of actual systems," Izv. AN SSSR, OTN No. 4-5 (1944).

72. G. Tammann. Metallography (Chemistry and Physics of Metals and Their Alloys) [Russian translation] (GNTI, 1931).

73. G. M. Kuznetsov. "Investigation of the processes of modification of the structure of binary alloys by the introduction of additions," Candidate Dissertation [in Russian] (Mintsvetmetzoloto, 1955).

74. A. A. Bochvar and V. V. Kuzina. "Influence of the character of crystallization and crystallization range on the mobility of liquid metal between growing crystals," Izv. AN SSSR, OTN No. 10 (1946).

75. R. W. Ruddle. "A preliminary study of the solidification of castings," Journ. Inst. Metals $\underline{77}$, 1 (1950).

76. R. W. Ruddle and A. L. Mincher. "The influence of alloy constitution on the mode of solidification of sand castings," Inst. Met. $\underline{79}$, 493 (1951).

77. K. Iwase, N. Asata, and N. Nasu. "On the nature of the peritectic reaction and the mechanism of the grain refinement resulting therefrom," Anniversary volume dedicated to K. Honda, Sednai, 1936: Journ. Chem. Soc. Japan $\underline{57}$, 310 (1936).

78. M. V. Mal'tsev. "Modification of the structures of metal alloys," Collection: Aluminum Alloys [in Russian] (Oborongiz, 1955).

79. L. L. Kulin. Surface Phenomena in Metals [in Russian] (Metallurgizdat, 1955).

80. Yu. A. Nekhendzi. Steel Castings [in Russian] (Metallurgizdat, 1948).

81. A. J. Murphy. Nonferrous Foundry Metallurgy. The Science of Melting and Casting Nonferrous Metals and Alloys (London, 1954).

82. W. C. Newell. The Casting of Steel (London - New York, 1955).

83. A. M. Korol'kov. "Shrinkage phenomena in alloys and the formation of cracks on solidification of metals [in Russian] (Mashgiz, 1958).

84. A. A. Bochvar. "Useful effect of the crystallization of alloys under pressure as a function of the composition of the alloy," Izv. AN SSSR, OTN No. 7 (1940).

85. A. M. Korol'kov and E. N. Timokhina. "The hydraulic resistance of Al–Si and Al–Cu alloys," Izv. AN SSSR, OTN No. 5 - 6 (1943).

86. A. A. Bochvar and Z. A. Sviderskaya. "Dependence of the imperviousness of castings on crystallization range," Izv. AN SSSR, OTN No. 11 - 12 (1943).

87. A. M. Korol'kov. "Surface tension and fluidity of aluminum- and zinc-based alloys," Collection: Experimental Techniques and Methods of High-Temperature Measurements [in Russian] (Izd. AN SSSR, 1959).

88. E. Gebhardt et al. "Die innere Reibung von flüssigen Al und Al-Legierungen. Z. Metallkunde $\underline{44}$, No. 11 (1953)

89. E. Gebhardt et al. "Die innere Reibung flüssiger Mg-Sn Legierungen," Z. Metallkunde $\underline{46}$, No. 9 (1955).

90. W. R. D. Jones and W. L. Bartlett. "The viscosity of aluminum and binary aluminum alloys," Journ. Inst. Metals $\underline{81}$, 145 (1952 - 53).

91. W. R. D. Jones and W. L. Bartlett. "The viscosity of copper and some binary copper alloys," Journ. Inst. Metals $\underline{83}$, 59 (1954 - 55).

92. B. A. Arbuzov and L. M. Guzhevina. "The parachor and structure of aromatic amines," ZhFKh $\underline{23}$, No. 9 (1949)

93. A. V. Rakovskii. Introduction to Physical Chemistry [in Russian] (Gonti, 1938).

94. H. K. Adam. The Physics and Chemistry of Surfaces [Russian translation] (Gostekhizdat, 1947).

95. S. D. Gromakov. "Determination of surface tension of liquids in a volume of millionths of a Milliliter," ZhFK $\underline{27}$, No. 4 (1953).

96. Ya. G. Dorfman. "On the theory of the surface tension of metals," Dokl. AN SSSR $\underline{41}$, No. 9, 386 (1943).

97. P. P. Pugachevich. "Experimental studies of the surface tension of metallic solutions," ZhFKh $\underline{25}$, No. 11 (1951)

98. P. P. Pugachevich and O. A. Timofeicheva. "Experimental investigation of the surface tension of sodium amalgams," Dokl. AN SSSR $\underline{94}$, No. 2 (1954).

99. P. P. Pugachevich and I. P. Altynov. "Temperature dependence of the surface tension of bismuth and its sodium and potassium alloys," Dokl. AN SSSR, 86, No. 1 (1953).

100. A. E. Samoilovich. "Experimental theory of the surface tension of metals," Dokl. AN SSSR 46, No. 9 (1945); ZhETF 16, 135 (1946).

101. A. E. Samoilovich. "On the question of the surface tension of metals," ZhFKh 21, No. 2 (1947).

102. A. E. Glauberman. "Theory of the surface tension of metals," ZhFKh 23, No. 2 (1949).

103. S. N. Zadumkin. "Connection between the surface tension of metals and their atomic volume," ZhETF 24, No. 5 (1953).

104. L. L. Kunin. "On the formulas for calculating the surface tension of metals," Dokl. AN SSSR 79, No. 1, 93 (1951).

105. N. A. Trifonov and G. K. Aleksandrov. "On the question of the application of surface tension in the physico-chemical analysis of rational systems," Izv. SFKhA 12, 85 (1940).

106. N. A. Trifonov. "On the form of the surface tension isotherms of binary liquid systems," Izv. SFKhA 12, 103 (1940).

107. N. A. Trifonov and A. G. Kholezova. "Surface tension of rational systems," Izv. SFKhA 12 (1940).

108. N. A. Trifonov and R. V. Merulin. "Surface tension of irrational systems," Izv. SFKhA 12, 139 (1940).

109. Kh. L. Strelets, A. Yu. Taits, and B. S. Gulyanitskii. Metallurgy of Magnesium [in Russian] (Metallurgizdat, 1950).

110. V. A. Kuznetsov, V. V. Ashpuri, and T. S. Troshina. "Study of the surface tension of thallium amalgams in a vacuum," Dokl. AN SSSR 101, No. 2, 301 (1955).

111. A. M. Korol'kov. "Surface tension of aluminum and its alloys," Izv. AN SSSR, OTN No. 2 (1956).

112. B. Ya. Pines. "Adsorption, surface tension and energy of mixing of binary metal alloys," ZhTF 22, No. 12 (1952).

113. A. E. Glauberman and A. M. Muzyrchuk. "Surface tension of binary metal alloys with body-centered and face-centered lattices," ZhFKh 38, No. 9, 1615 (1954).

114. J. W. Taylor. "The surface tensions of liquid metal solutions," Acta Metallurg. 4, IX, 460 (1956).

115. F. Sauerwald. Textbook on the Science of Metals [Russian translation] (ONTI, 1932).

116. F. Sauerwald. Z. f. anorg. allgem. Chem. 181, 347 (1929); 181, 353 (1929).

117. H. T. Greenaway. "The surface tension and density of lead-antimony and cadmium-antimony alloys," J. Inst. Metals 74, 133 (1948).

118. E. Pelzel. Berg-Hüttenmännische Monatshefte 93, 248 (1948); 94, 10 (1949).

119. V. V. Bakridze and B. Ya. Pines. "Surface tension of Pb–Sn, Bi–Pb, Bi–Cd binary alloys," ZhTF 23, No. 9 (1953).

120. B. V. Stark and S. I. Filippov. "Adsorption phenomena on the surface of liquid steel," Izv. AN SSSR, OTN No. 3 (1949).

121. T. P. Kolesnikova and A. M. Samarin. "Influence of manganese, chromium, and vanadium on the surface tension of liquid iron," Izv. AN SSSR, OTN No. 5 (1956).

122. T. F. Bas and G. G. Kellog. "Influence of dissolved sulfur on the surface tension of liquid copper," Problemy sovremennoi metallurgii No. 2 (1945).

123. P. A. Rebinder and Z. S. Lipman. "Physicochemical principles of the modification of metals and alloys by small surface-active additions," Collection: Research in the Field of Applied Physical Chemistry of Surface Phenomena [in Russian] (ONTI, 1936).

124. Yu. A. Klyachko et al. "Influence of boron on the surface tension of steel," Dokl. AN SSSR 72, 45, 927 (1950).

125. S. M. Baranov. "A surface-active constituent and its influence on the properties of steel," Dissertation for Doctor's Degree [in Russian] AN SSSR, (1955).

126. M. V. Mal'tsev. "Modification of the structure of nonferrous metals and alloys," Tsent. inst. informatsii tsvetnoi metallurgii, Bull. No. 9 (62) (1956).

127. G. M. Kuznetsov. "Influence of vibration on the crystallization of modified aluminum-silicon alloys," Dokl. AN SSSR 101, No. 1 (1956).

128. D. Paul and E. Shile. Measurement of the Surface Tension of Cast Iron. XXIII International Congress of Foundrymen [Russian translation] (Mashgiz, 1958).

129. O. S. Bobkova and A. M. Samarin. "Connection between the surface tension of chrome-nickel alloys and some of the properties of chrome-nickel alloys," Izv. AN SSSR, OTN No. 2 (1954).

130. V. G. Zhivov. "Determination of the surface tension of molten Al, Mg, Na, and K," Tr. VAMI No. 14 (1937).

131. S. W. Smith. "The surface tension of molten metals," Journ. Inst. Metals $\underline{12}$, 168 (1914).

132. Yu. A. Klyachko. "Measurement of surface tension as a method of technological characterization," Zav. lab. $\underline{6}$, 1376 (1937).

133. S. V. Sergeev. Physicochemical Properties of Liquid Metals [in Russian] (Oborongiz, 1952).

134. A. P. Smiryagin. Industrial Nonferrous Metals and Alloys [in Russian] (Metallurgizdat, 1956).

135. Liquid-Metals Handbook (Atomic Energy Commission, Washington, 1952).

136. A. Portevin and P. Bastien. "La résistance mécanique de la peau d' alumine et son influence sur la tension superficielle du métal fondu," C. R. $\underline{202}$, 1072 (1936).

137. Yu. A. Klyachko. "Measurement of surface tension as a method of technological characterization," Zav. lab. $\underline{6}$, 1376 (1937).

138. A. I. Belyaev and E. A. Zhemchuzhina. Surface Phenomena in Metallurgical Processes [in Russian] (Metallurgizdat, 1952).

139. S. I. Filippov. Theory of the Steel Decarburization Process [in Russian] (Metallurgizdat, 1956).

140. S. A. Voznesenksii and P. A. Rebinder. Guide to Laboratory Work in Physical Chemistry [in Russian] (GIZ, 1928).

141. V. K. Semenchenko. "The molecular theory of adsorption in solutions," Izv. SFKhA $\underline{21}$, 16 (1952).

142. K. I. Portnoi and A. A. Lebdev. Magnesium Alloys [in Russian] (Metallurgizdat, 1952).

143. A. S. Lugaskov. Magnesium Alloy Castings [in Russian] (Oborongiz, 1942).

144. A. P. Pronov. "Crystallization and properties of liquid steel in continuous casting conditions," Collection: Continuous Casting of Steel [in Russian] (Izd. AN SSSR, 1956).

145. S. M. Voronov. "Processes of melting magnesium alloys," Collection: Casting of Magnesium Alloys [in Russian] (Oborongiz, 1952).

146. A. M. Korol'kov and E. S. Kadaner. Fluidity of Metals and Alloys [in Russian] (Metallurgizdat, 1952).

147. A. M. Korol'kov. "Flow of metals and alloys in channels," Transactions of Conference on the Hydrodynamics of Molten Metals [in Russian] (Izd. AN SSSR, 1958).

148. "Determination of fluidity of aluminum and magnesium casting alloys," Standard SMI [in Russian] (Oborongiz, 1954) pp. 215 - 254.

149. Yu. A. Nekhendzi and A. M. Samarin. "Fluidity of steel as technological test for the assessment of its quality," Transactions of First Conference on the Physicochemical Principles of Steel Production [in Russian] (1951) p. 423.

150. Yu. A. Nekhendzi. "Fluidity and quality of castings," Collection: Recent Developments in the Theory and Practice of Foundry Production [in Russian] (Mashgiz, 1956).

151. A. Portevin and P. Bastien. Journ. Inst. Metals $\underline{54}$, 45 (1934).

152. A. A. Bochvar and I. I. Novikov. "The solid-liquid state of alloys in the period of their crystallization," Collection: Technology of Nonferrous Metals [in Russian] (Metallurgizdat, 1952).

153. N. N. Kurnakov, N. N. Sirota, and M. Ya. Troneva. "The influence of silicon and other elements on the fluidity of blast-furnace ferrochrome," Dokl. AN SSSR $\underline{51}$, No. 3 (1946).

154. N. N. Kurnakov and M. Ya. Troneva. "On the determination of fluidity and viscosity of metal alloys," Dokl. AN SSSR $\underline{51}$, No. 5 (1946).

155. N. N. Kurnakov and M. Ya. Troneva. "On the fluidity of iron alloys," Izv. SFKhA $\underline{20}$ (1951).

156. V. Kondic. Journ. Inst. Metals $\underline{75}$ (1940).

157. K. V. Peredel'skii. Casting of Nonferrous Alloys in Metal Molds [in Russian] (Mashgiz, 1951).

158. K. I. Vashchenko. Chemically Resistant Castings [in Russian] (Mashgiz, 1946).

159. T. P. Yao and V. Kondic. "What is fluidity?" Met. Ind. $\underline{79}$, 435 (1951).

160. Metals Handbook (Amer. Soc. Metals, 1948).

161. J. Czikel and T. Grossmann. Abstract in the journal Metallurgiya, No. 3, 2230 (1956).

162. A. A. Semionov. Typographical Alloys [in Russian] (Gizlegprom, 1941).

163. Metallurgist's Reference Book for Nonferrous Metals [in Russian] (Metallurgizdat, 1953) Vol. I.

164. E. Gebhardt. "Innere Reibung Al–Zn," Z. Metallkunde $\underline{45}$, No. 2 (1945).

165. M. R. Kryanin and G. V. Suskho. Comparative Properties of Steel for Castings Melted by the Basic and Acid Processes [in Russian] (Mashgiz, 1951).

166. V. I. Prosvirin, V. S. Rakovskii, and A. F. Silaev. "Production of metallic powders by atomization of liquid alloys," Vestnik mashinostroeniya No. 7 (1951).

167. A. P. Smiryagin. Industrial Nonferrous Metals and Alloys [in Russian] (Metallurgizdat, 1949).

168. E. N. Kulagina. "Refinement of aluminum alloys and quality control by structure-specimen," Collection: Nonferrous castings [in Russian] (Mashgiz, 1954).

169. E. Elliott. "The 1955 specification revision," Light Metals 18, 409 (1955).

170. V. I. Dobatkin. Continuous Casting and the Casting Properties of Alloys [in Russian] (Oborongiz, 1948).

171. Continuous Casting of Steel Collection [Russian translation] (INI AN SSSR, 1955).

172. P. N. Bidulya. Foundry Production [in Russian] (Metallurgizdat, 1955).

173. Turner, Murray, Haughton and others. Journ. Inst. Metals 1909, 1911, 1913.

174. G. Sachs. Practical Metallurgy [Russian translation] (ONTI, 1926) Vol. I.

175. A. A. Bochvar, Z. A. Sviderskaya, and E. K. Korbut. "On the question of the expansion of some alloys on crystallization," Izv. AN SSSR, OTN No. 4 (1947).

176. A. A. Bochvar and O. S. Zhadaeva. "Theory of shrinkage phenomena in alloys," Liteinoe delo No. 5 (1941).

177. A. A. Bochvar and V. I. Dobatkin. "The temperature curve of the commencement of linear shrinkage in binary alloys," Izv. AN SSSR, OTN No. 1-2 (1945).

178. A. A. Bochvar and Z. A. Sviderskaya. "The destruction of castings by the action of shrinkage stresses in the crystallization period as a function of composition," Izv. AN SSSR, OTN No. 3 (1947).

179. Encyclopedia of the Physics of Metals [in Russian] (ONTI, 1937) Vol. I.

180. A. M. Korol'kov. "On the estimation of shrinkage in metals," Zav. lab. 1 (1947).

181. J. Leuser. "Eine einfache Methode zur Bestimmung der Metallschwindung bei Güssen," Metallwirtschaft 19, No. 5, 77, 140.

182. O. Bauer and W. Heidenheim. "Verhalten der Al-Zn Legierungen," Z. Metallkunde 16, 221 (1924).

183. A. Burkhard. The Mechanical and Technological Properties of Pure Metals [Russian Translation] (Metallurgizdat, 1941).

184. Landolt. Phys. Chem. Tab. (Berlin, 1936) Vol. III.

185. A. Ledebur-Bauer. Die Legierungen (Berlin, 1924).

186. Technical Encyclopedia. Reference Book [in Russian] (ONTI, 1929) Vol. 2.

187. A. M. Korol'kov. "Shrinkage of copper alloys," Tsvetnye metally No. 3 (1949).

188. A. M. Korol'kov and E. S. Kadaner. "Anomalous cases of the variation of linear shrinkage of alloys on variation in their composition," Collection: Research on Nonferrous Metal Alloys [in Russian] (Izd. AN SSSR, 1955) Vol. I.

189. A. M. Korol'kov. Shrinkage Phenomena in Alloys and Cracking on Solidification [in Russian] (Izd. AN SSSR, 1957).

190. N. I. Novikov. "Investigation of alloys for hardness and bending near the solidus," Collection: Technology of Nonferrous Metals [in Russian] (Metallurgizdat, 1952).

191. A. Beck. Magnesium and Its Alloys [Russian translation] (Oborongiz, 1941).

192. P. Hindnert and H. Krider. "Thermal expansion of aluminum and some aluminun alloys," Research of the National Bureau of Standards 48, No. 3 (1952).

193. M. G. Oknov. "Variation in Volume of metals on quenching," ZhRMO No. 3 (1915).

194. V. F. Gachkovskii and P. T. Strelkov. ZhETF 7, No. 4 (1937).

195. L. Fromer. Casting under Pressure [Russian translation] (ONTI, INTPI, 1935).

196. A. P. Pronov. Shrinkage and Strength of Steel during and After Solidification [in Russian] (Metallurgizdat, 1950).

197. A. A. Ryzhikov. Improving the Quality of Castings [in Russian] (Mashgiz, 1952).

198. N. E. Chernobaev. Casting in Chills [in Russian] (Mashgiz, 1947).

199. D. K. Butakov. "Method of assessing the susceptibility of steel to hot cracking," Vestnik mashinostroeniya No. 12 (1950).

200. J. Verö. Metal Industry 48, No. 15, 17, 431, 494 (1936).

201. V. A. Livanov. "Casting of large ingots for the production of aluminum alloy sheets," Collection: Aluminum Alloys [in Russian] (Oborongiz, 1955).

202. I. N. Fridlyander, F. V. Tulyankin et al. "The causes of the occurrence of cracks in flat ingots of the alloy V-95," Collection: Aluminum Alloys [in Russian] (Oborongiz, 1955).

203. B. I. Medovar. "On the question of the nature of hot cracks in welds," Avtomaticheskaya svarka No. 4 (1954).

204. N. F. Lashko and S. V. Lashko-Avakyan. Metallurgy of Welding [in Russian] (Mashgiz, 1954).

205. N. N. Prokhorov. Hot Cracks in Welding [in Russian] (Mashgiz, 1952).

206. A. I. Veinik. Thermal Principles of the Theory of Casting [in Russian] (GNTI, 1953).

207. A. A. Bochvar and Khakimdzhanova. "The susceptibility of aluminum alloys to cracking under the action of shrinkage stresses," Metallurg No. 2 (1939).

208. A. A. Bochvar. "On the question of the optimum content of eutectic in casting alloys," Izv. AN SSSR, OTN No. 6 (1944).

209. A. M. Korol'kov and E. S. Kadaner. "Application of the microhardness method to the study of the structural constituents of nonferrous alloys," Collection: Microhardness [in Russian] (Izd. AN SSSR, 1951).

210. W. J. Pumphry and P. H. Jennings. "A consideration of the nature of the brittleness at temperatures above the solidus in castings and welds in aluminum alloys," Journ. Inst. Metals. 75, 235 (1948 - 49).

211. P. H. Jennings, A. R. Singer, and W. J. Pumphry. "Hot-shortness of some high-purity alloys in the system Al–Cu–Si and Al–Mg–Si," Journ. Inst. Metals 74, 227 (1948).

212. P. H. Jennings and W. J. Pumphry. "A consideration of the constitution of the aluminum-iron-silicon alloys and its relation to cracking above the solidus," Journ. Inst. Metals 74, 249 (1948).

213. W. J. Pumphry and D. C. Moore. "Cracking during and after solidification in some aluminum-copper-magnesium alloys of high purity," Journ. Inst. Metals 74, 425 (1948).

214. L. M. Postnova and B. B. Gulyaev. "Investigation of the mechanical properties of steel in the solidification period and analysis of the process of hot cracking in continuous casting," Collection: Continuous Casting of Steel [in Russian] (Izd. AN SSSR, 1956).

215. V. A. Éfimov, V. I. Danilin, and M. P. Lapshova. "Influence of the crystallization conditions of steel on ingot rejection due to cracking," Stal' No. 7 (1955).

216. V. N. Saveiko. "The mechanism of hot cracking in steel castings," Collection: Continuous Casting of Steel [in Russian] (Izd. AN SSSR, 1956).

217. V. G. Gruzin. "Pouring temperature of metal and hot cracking," Collection: Continuous Casting of Steel [in Russian] (Izd. AN SSSR, 1956).

218. A. A. Bochvar. "Prospects of discovering new casting alloys," Izv. AN SSSR, OTN No. 9 (1942).

219. A. G. Spasskii. "Methods of economizing in nonferrous metals in foundry practice and the immediate problems in the field of research work," Collection: Nonferrous Castings [in Russian] (Mashgiz, 1954).

220. S. G. Glazunov and S. I. Spektorova. Technological Properties of Aluminum and Magnesium Casting Alloys [in Russian] (Oborongiz, 1950).

221. I. F. Kolobnev. "Aluminum casting alloys," Collection: Aluminum Alloy Castings [in Russian] (Mashgiz, 1953).

222. A. P. Belyaev et al. "Aluminum (secondary) casting alloys with zinc, silicon, and copper," Collection: Treatment of Nonferrous Metals and Alloys [in Russian] (Metallurgizdat, 1953).

223. A. M. Osokin. "Practical measures for combating hot-shortness in magnesium chill-cast alloys," Liteinoe proizvodstvo No. 1 (1956).

224. A. A. Ryzhnikov. Shrinkage Heads for Steel Castings [in Russian] (Mashgiz, 1947).

225. V. N. Maslova. "Choice of type of mold for open-hearth rail steel," Collection: The Steel Ingot [in Russian] (Metallurgizdat, 1952).

226. K. I. Akimova. "Shrinkage phenomena in alloys as a function of composition," Collected Works of Mintsvetzoloto [in Russian] No. 8 (Metallurgizdat, 1940).

227. A. M. Korol'kov. "Hydraulic resistance of cast copper," Izv. AN SSSR, OTN No. 1 (1948).

228. Slavinskii et al. "Nature of silicon bronzes and their utilization for castings," Metallurg No. 1 (1935).

229. V. V. Krymov. Casting of Magnesium Alloys [in Russian] (Oborongiz, 1945).

230. M. V. Sharov. Aluminum and Magnesium Casting Alloys [in Russian] (Mashgiz, 1951).

231. I. F. Kolobnev, M. B. Al'tman, and O. B. Lotareva. "The nature of black fracture in aluminum-magnesium alloy castings," Tr. VIAM [in Russian] (Oborongiz, 1948) Vol. 2.

232. A. A. Bochvar. Porosity in Nonferrous Castings and Methods of Combating It [in Russian] (Mashgiz, 1941).

233. I. F. Kolobnev and M. A. Al'tman. Gas Porosity and Methods of Combating It in Aluminum Alloys [in Russian] (Moscow, 1948).

234. N. N. Beloysov and V. A. Egorova. "Improving the properties of castings of the alloy AL-8," Collection: Recent Developments in the Theory and Practice of Foundry Production [in Russian] (Mashgiz, 1956).

235. I. N. Golikov. Dendritic Segregation in Steel [in Russian] (Metallurgizdat, 1958).

236. M. V. Chukhrov. "Study of the crystallization processes in magnesium alloy ingots," Collection: Solidification of Metals [in Russian] (Mashgiz, 1958).

237. B. B. Gulyaev and O. M. Magnitskii. "Study of the influence of alloy composition on the kinetics of solidification," Collection: Solidification of Metals [in Russian] (Mashgiz, 1958).

238. W. T. Read. Dislocations in Crystals [Russian translation] (Metallurgizdat, 1957).